CARTOON PHYSICS

A GRAPHIC NOVEL GUIDE
TO SOLVING PHYSICS PROBLEMS

BY SCOTT CALVIN AND KIRIN EMLET FURST

Additional artwork by Elena Hartley

CRC Press
Taylor & Francis Group
Boca Raton London New York

CRC Press is an imprint of the
Taylor & Francis Group, an **informa** business

First edition published 2022
by CRC Press
6000 Broken Sound Parkway NW, Suite 300, Boca Raton, FL 33487-2742

and by CRC Press
4 Park Square, Milton Park, Abingdon, Oxon, OX14 4RN

ISBN: 9781032210414 (hbk)
ISBN: 9781138598782 (pbk)
ISBN: 9780429486128 (ebk)

DOI: 10.1201/9780429486128

Publisher's note: This book has been prepared from camera-ready copy provided by the authors.

TABLE OF CONTENTS

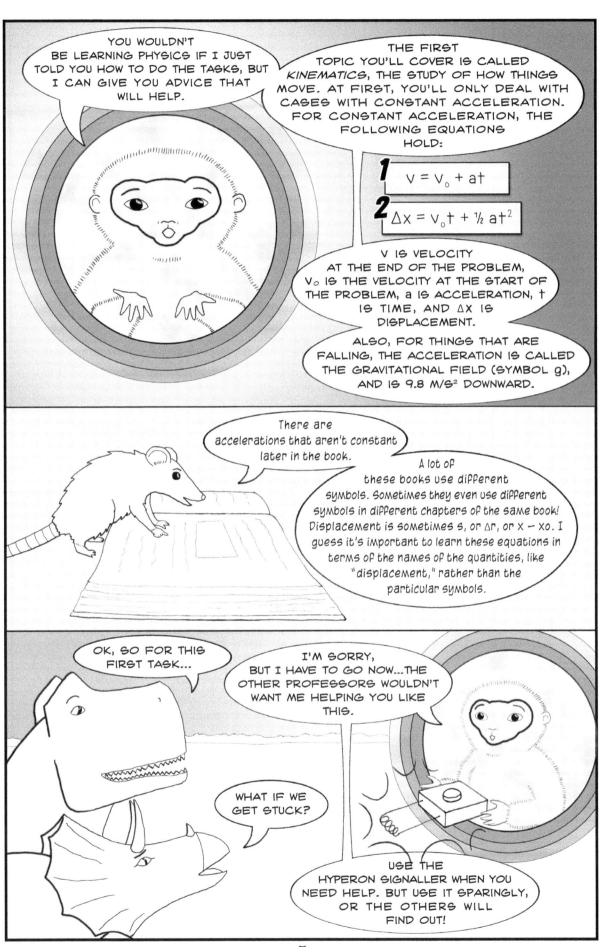

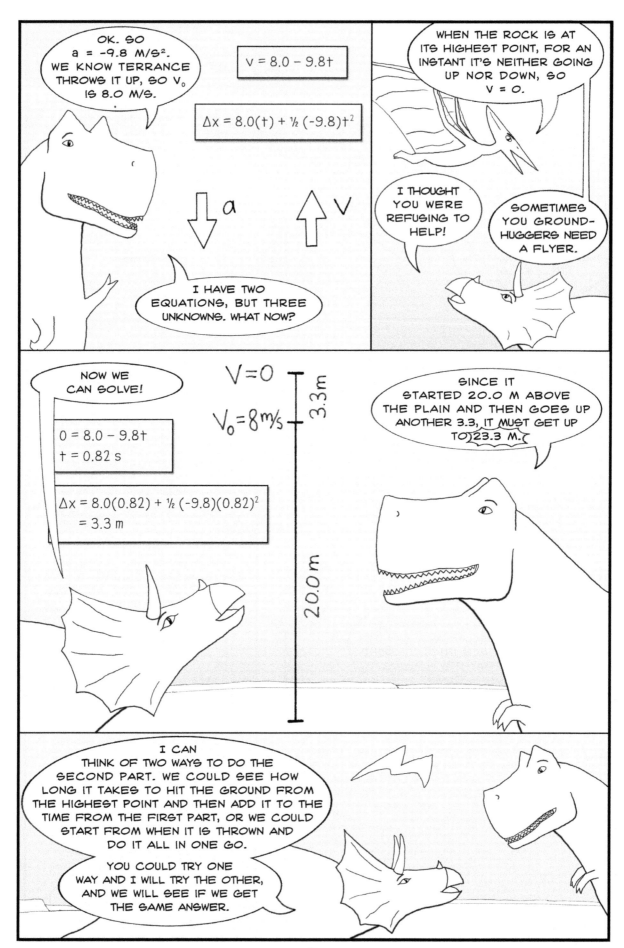

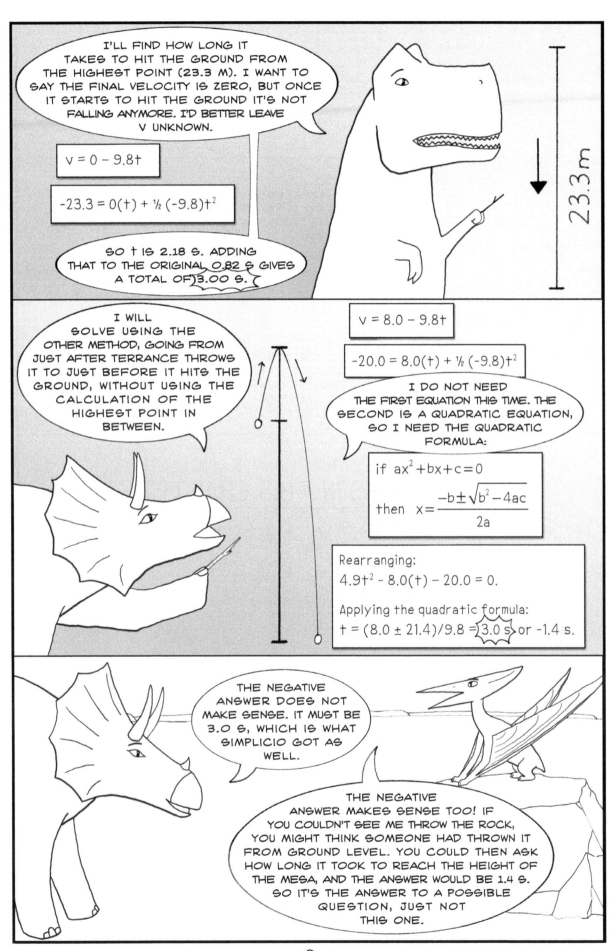

EQUATIONS IN THIS CHAPTER

$$v = v_0 + at$$
$$\Delta x = v_0 t + \tfrac{1}{2} at^2$$

WORDS OF WISDOM

"Initial" and "Final" are when you start and stop paying attention

If you have as many equations as unknown variables, it's possible to solve

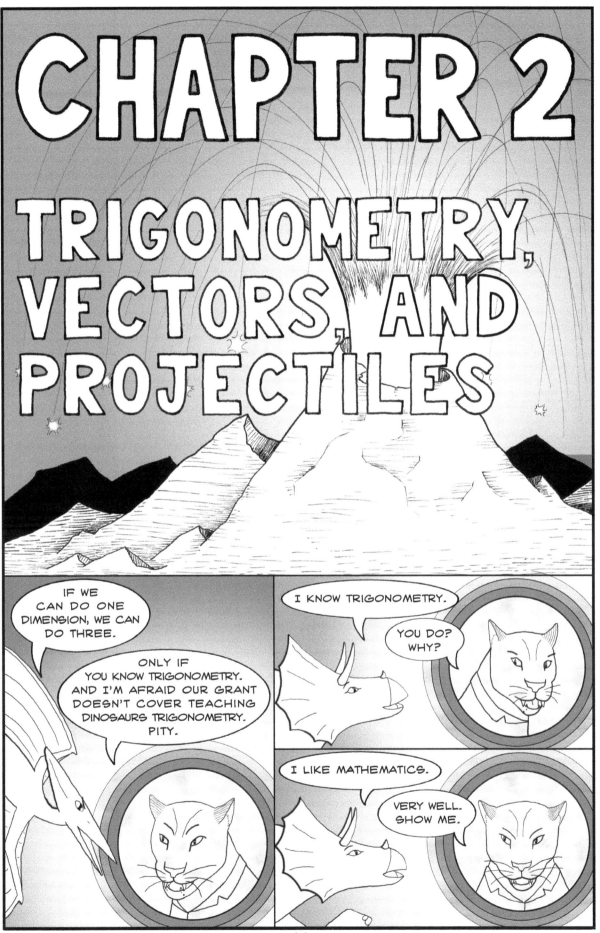

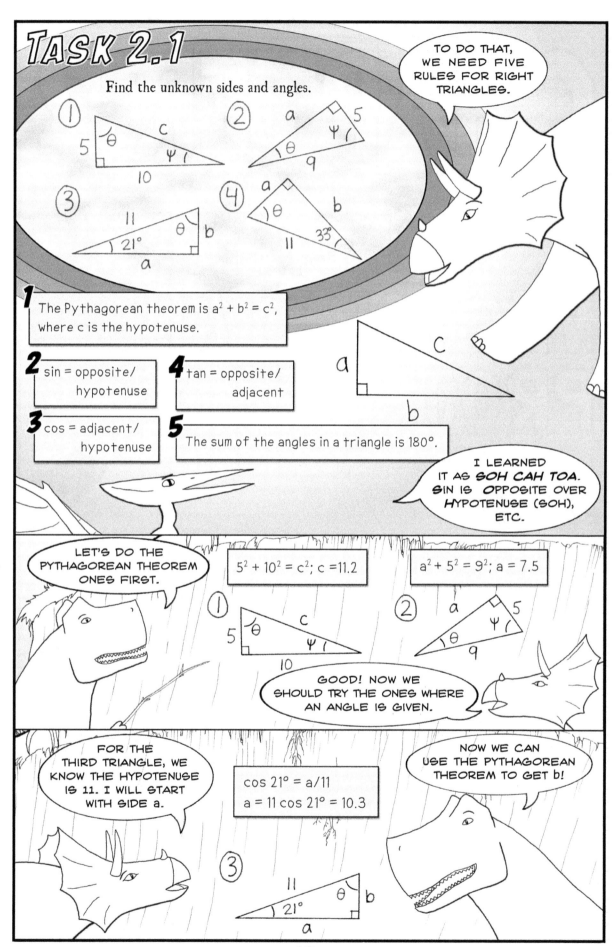

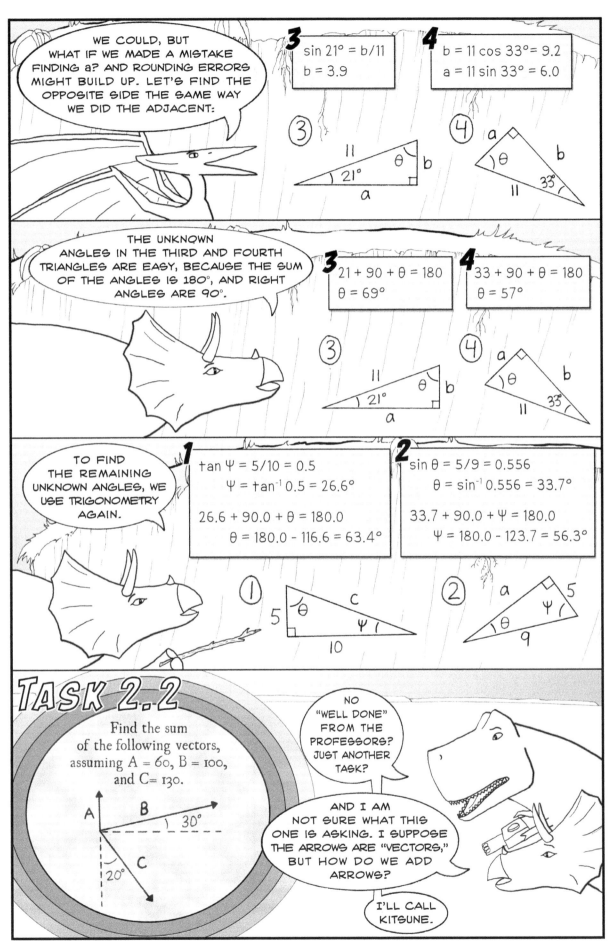

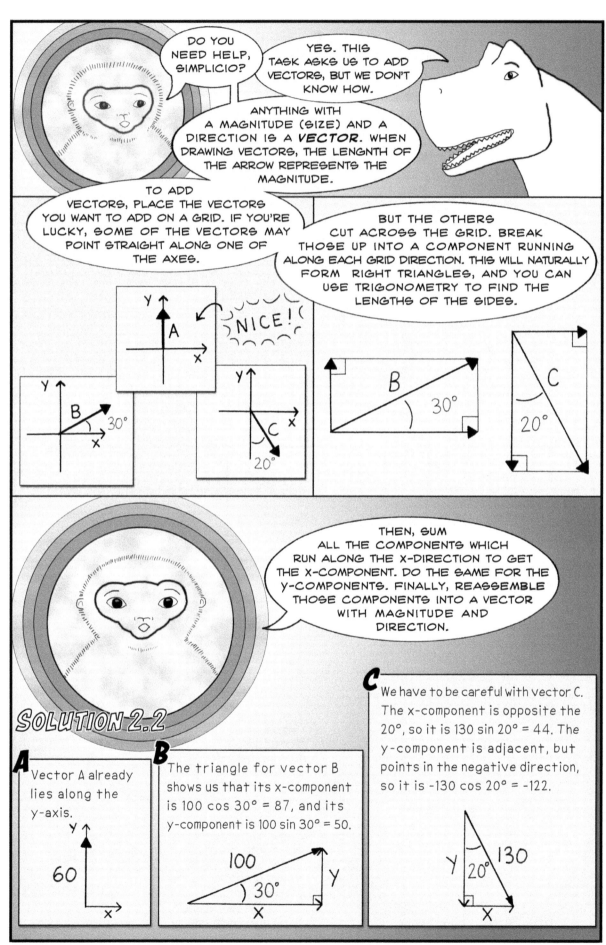

14

Adding them gives us 87 + 44 = 131 in the x-direction, and 60 + 50 + (-122) = -12 in the y-direction.

WHAT ABOUT THE MINUS SIGN?

WE HAVE ALREADY SHOWN THE DIRECTION OF THE VECTOR IN OUR SKETCH. THAT SHOULD BE GOOD ENOUGH.

The hypotenuse is given by $131^2 + 12^2 = c^2$, so the magnitude is 132. The angle it makes with the x-axis is given by $\tan \theta = 12/131 = 0.093$, so $\theta = \tan^{-1} 0.093 = 5.3°$.

TASK 2.3

Terrance throws a rock off the edge of a 20.0 m tall vertical mesa with an initial speed of 8.0 m/s at an angle 30° above the horizontal. How far from the base of the mesa does it land? How high above the surrounding plain does it get? How long does it take (from when it was thrown) to hit the ground?

THIS IS SIMILAR TO ONE OF THE KINEMATICS TASKS, EXCEPT IT IS TWO-DIMENSIONAL.

LET'S START BY TAKING THE INITIAL VELOCITY AND BREAKING IT INTO COMPONENTS.

The x-component is 8.0 cos 30° = 6.9

The y-component is 8.0 sin 30° = 4.0

Careful! That's correct, but don't get in the habit of thinking x-components are always cosine and y-components are always sine. This x-component is only cosine because it's adjacent to the angle.

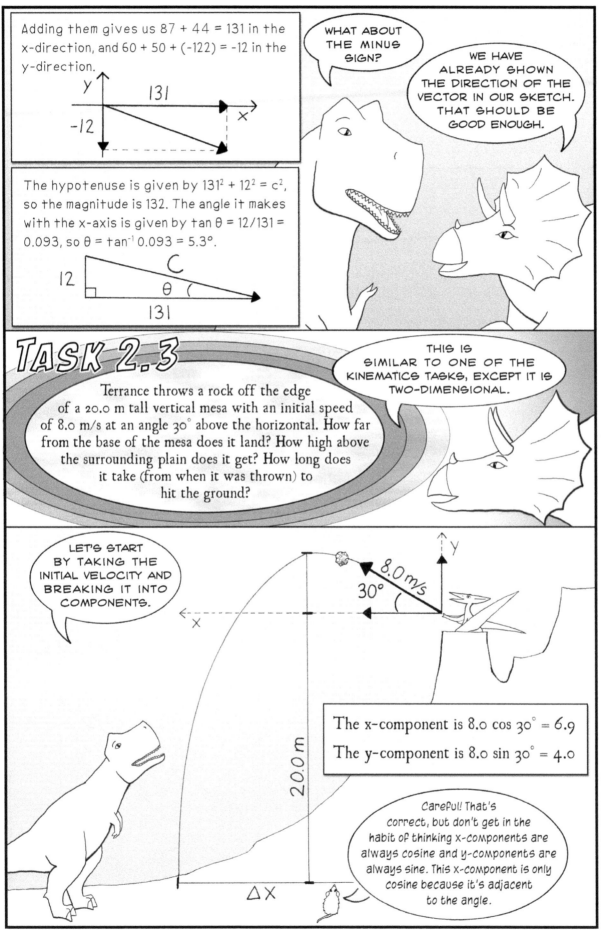

Now let's use our two kinematics equations with the x-component. The acceleration is downward, so it's zero in the x-direction.

$v_x = 6.9 + (0)t = 6.9$

$\Delta x = 6.9(t) + \frac{1}{2}(0)t^2 = 6.9t$

TECHNICALLY, WE HAVE FOUR EQUATIONS AND FOUR UNKNOWNS: v_x, Δx, v_y, AND t.

BUT WE KNOW v_x. THE FIRST EQUATION TELLS US IT'S 6.9.

THEN WE ONLY HAVE THREE EQUATIONS. EITHER WAY, WE'RE READY TO SOLVE.

The question asks several different things regarding y. For now, let's choose the final point to be just before it hits the plain, so that $\Delta y = -20.0$ m.

$v_y = 4.0 + (-9.8)t$

$\Delta y = -20.0 = 4.0(t) + \frac{1}{2}(-9.8)t^2$

Solving the quadratic for t gives 2.5 or -1.7 s. We want the positive one, 2.5 s, to answer the last part of the task.

We can also answer the first part:
$\Delta x = 6.9t = 6.9(2.5) = $ 17 m

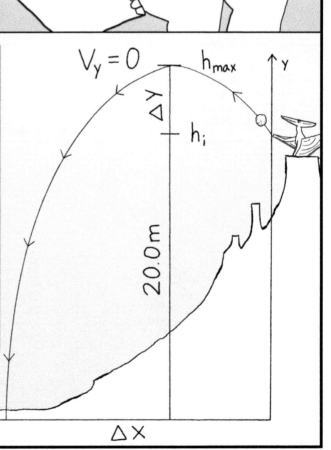

To figure out how high it goes, we need to choose a different final point. At its highest, it's not going up or down, so $v_y = 0$. Our four equations look like this:

$v_x = 6.9 + (0)t = 6.9$
$\Delta x = 6.9(t) + \frac{1}{2}(0)t^2 = 6.9t$
$v_y = 0 = 4.0 + (-9.8)t$
$\Delta y = 4.0(t) + \frac{1}{2}(-9.8)t^2$

The third gives $t = 0.41$ s. The last gives $\Delta y = 4.0(0.41) + \frac{1}{2}(-9.8)(0.41)^2 = 0.8$ m.

So it reaches 20.8 m above the surrounding plain.

$V_y = 0$ h_{max} y

ΔY h_i

20.0m

x

ΔX

16

EQUATIONS IN THIS CHAPTER

$$a^2 + b^2 = c^2$$

SOH CAH TOA

The sum of the angles in a triangle is 180°

WORDS OF WISDOM

Use cosine to find the component adjacent to the angle and sine to find the component opposite to it

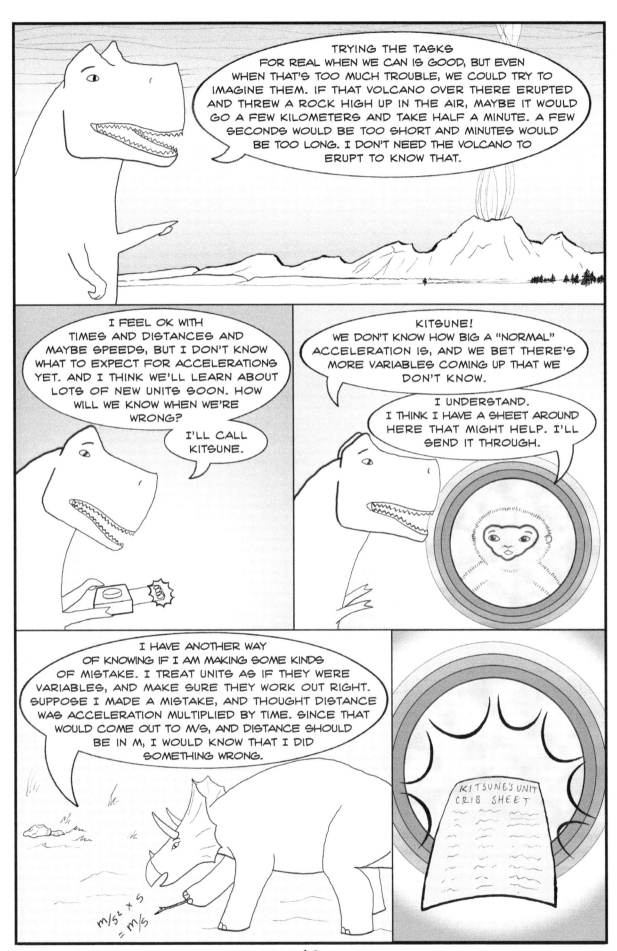

*Kitsune's Units Crib Sheet is reproduced on page 268 of this book.

WORDS OF WISDOM

Estimation, either by experiment or imagination, helps catch mistakes

Units can be treated like variables

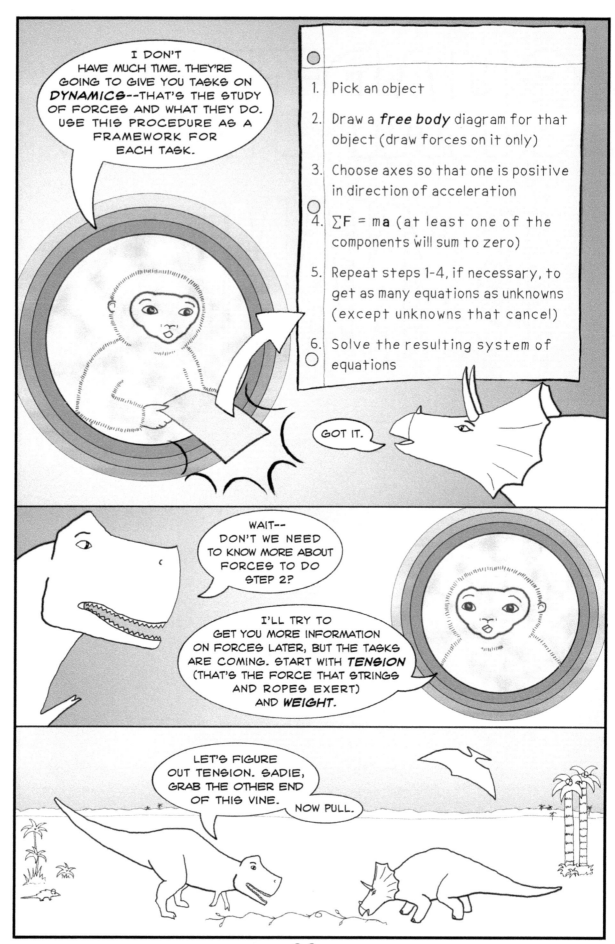

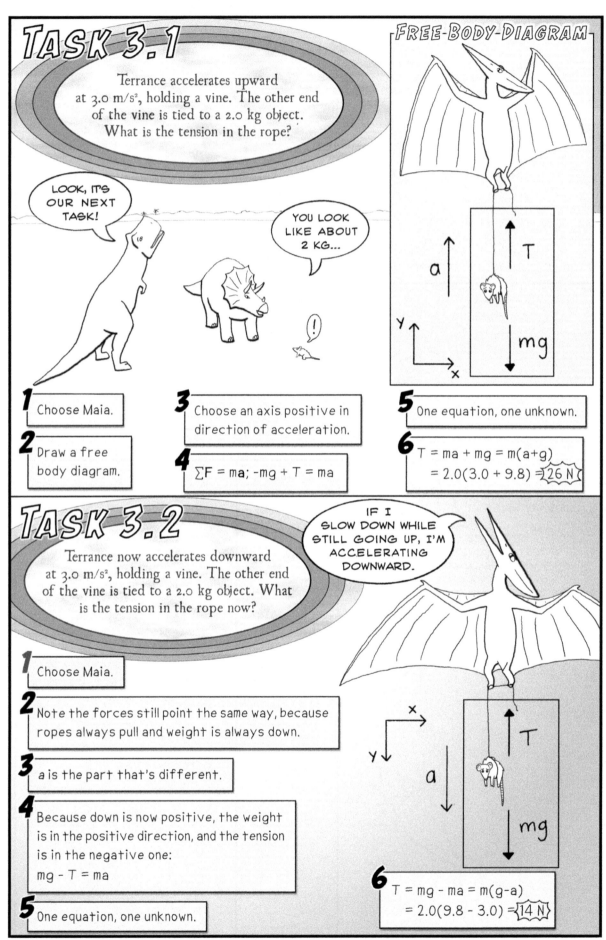

TASK 3.1

Terrance accelerates upward at 3.0 m/s², holding a vine. The other end of the **vine** is tied to a 2.0 kg object. What is the tension in the rope?

FREE-BODY-DIAGRAM

LOOK, IT'S OUR NEXT TASK!

YOU LOOK LIKE ABOUT 2 KG...

1 Choose Maia.

2 Draw a free body diagram.

3 Choose an axis positive in direction of acceleration.

4 $\sum F = ma$; $-mg + T = ma$

5 One equation, one unknown.

6 $T = ma + mg = m(a+g)$
$= 2.0(3.0 + 9.8) = 26 \text{ N}$

TASK 3.2

Terrance now accelerates downward at 3.0 m/s², holding a vine. The other end of the vine is tied to a 2.0 kg object. What is the tension in the rope now?

IF I SLOW DOWN WHILE STILL GOING UP, I'M ACCELERATING DOWNWARD.

1 Choose Maia.

2 Note the forces still point the same way, because ropes always pull and weight is always down.

3 a is the part that's different.

4 Because down is now positive, the weight is in the positive direction, and the tension is in the negative one:
$mg - T = ma$

5 One equation, one unknown.

6 $T = mg - ma = m(g-a)$
$= 2.0(9.8 - 3.0) = 14 \text{ N}$

24

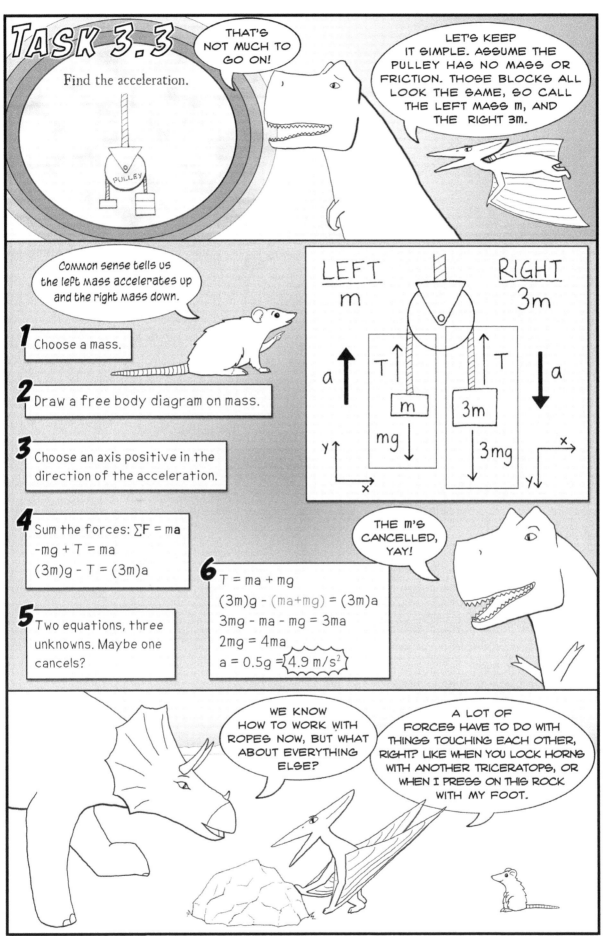

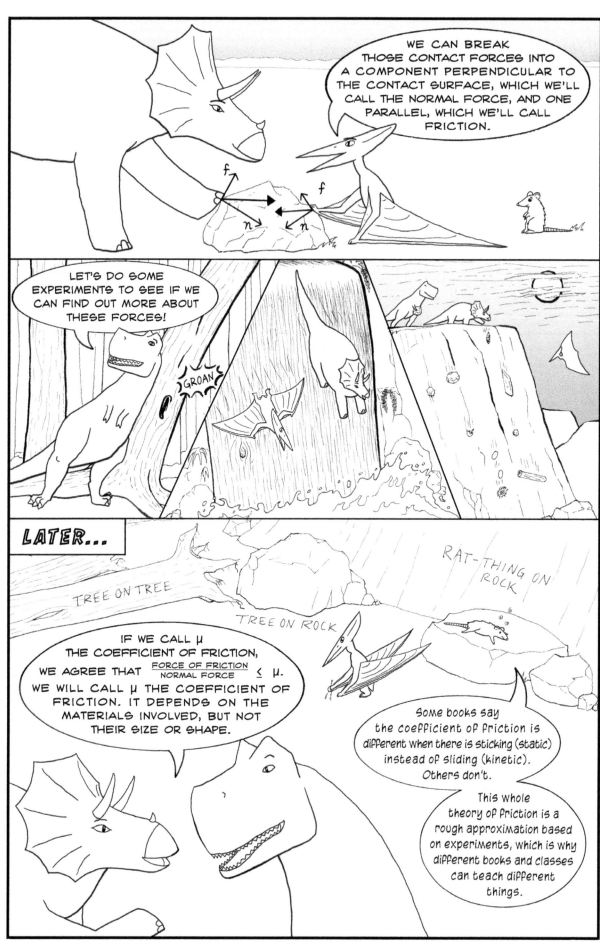

A 3.0 kg object is attached by a rope to a 10.0 kg object, which in turn is attached by another rope to a second 3.0 kg object. They rest on a horizontal surface and the coefficient of friction between each object and the surface is 0.400. The front object is pulled with a 150.0 N horizontal force. What is the acceleration of the system and the tension in each section of rope?

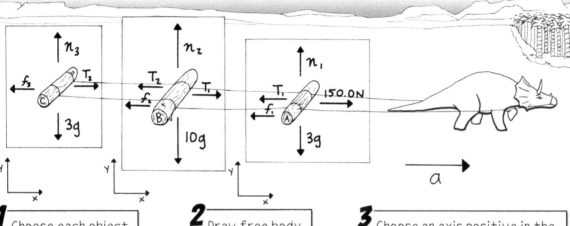

1 Choose each object individually.

2 Draw free body diagrams.

3 Choose an axis positive in the direction of the acceleration.

4 $\Sigma F = ma$

The experimental equation says <, but almost all problems involve the maximum friction for the set-up, so we use =.

	ΣF_x	ΣF_y	μ
A	$-f_1 - T_1 + 150.0 = (3.0)a$	$n_1 - (3.0)g = 0$	$f_1/n_1 = 0.400$
B	$-f_2 - T_2 + T_1 = (10.0)a$	$n_2 - (10.0)g = 0$	$f_2/n_2 = 0.400$
C	$-f_3 + T_2 = (3.0)a$	$n_3 - (3.0)g = 0$	$f_3/n_3 = 0.400$

5 Nine equations, nine unknowns!

6

ΣF_y:

$\quad n_1 = 3.0g \quad n_2 = 10.0g \quad n_3 = 3.0g$

μ:

$\quad f_1 = 1.20g \quad f_2 = 4.00g \quad f_3 = 1.20g$

ΣF_x:

$\quad$ A: $\quad -1.20g - T_1 + 150.0 = 3.0a$

$\quad$ B: $\quad -4.00g - T_2 + T_1 = 10.0a$

$\quad$ C: $\quad -1.20g + T_2 = 3.0a$

Rearrange to solve T_1 and T_2 in terms of a!

Solve for a:

$-4.00g - (3.0a + 1.20g) +$

$\quad (-1.20g + 150.0 - 3.0a) = 10.0a$

$-6.40g + 150.0 = 16.0a$

$-62.7 + 150.0 = 87.3 = 16.0a$

$a = 5.46 \text{ m/s}^2$

Solve for T_1 and T_2:

$T_1 = -1.20g + 150.0 - 3.0a = 122N$

$T_2 = 3.0a + 1.20g = 28N$

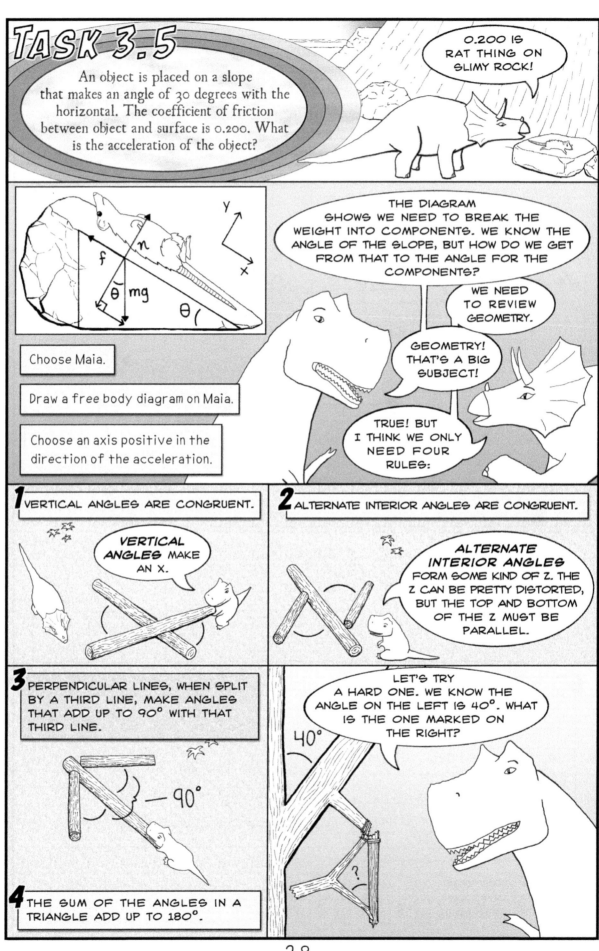

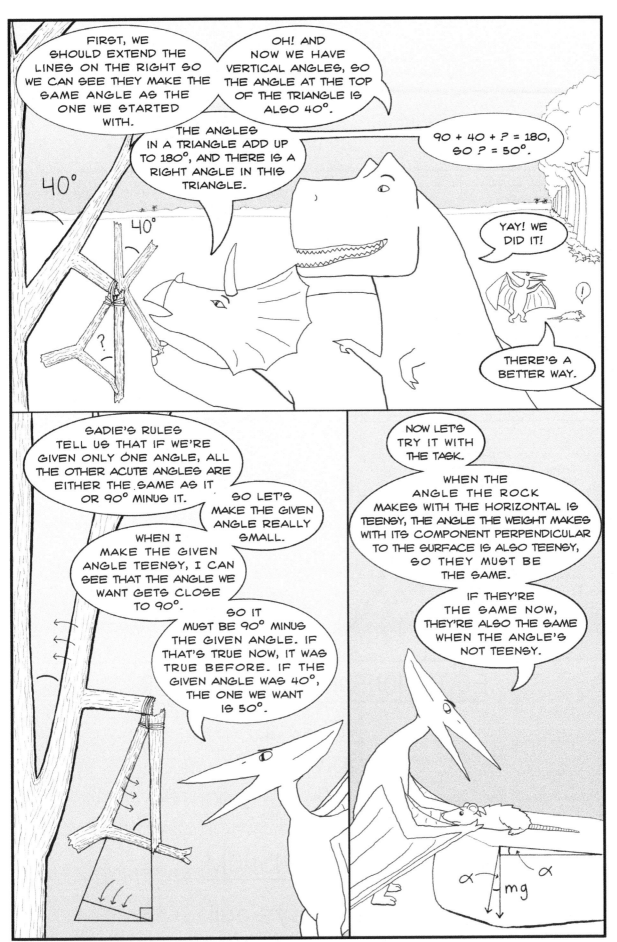

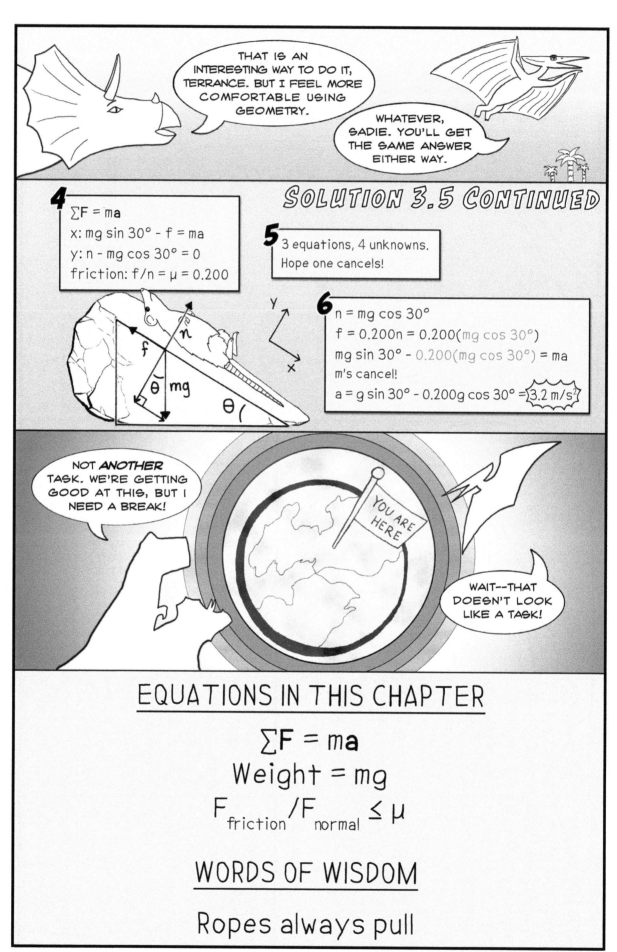

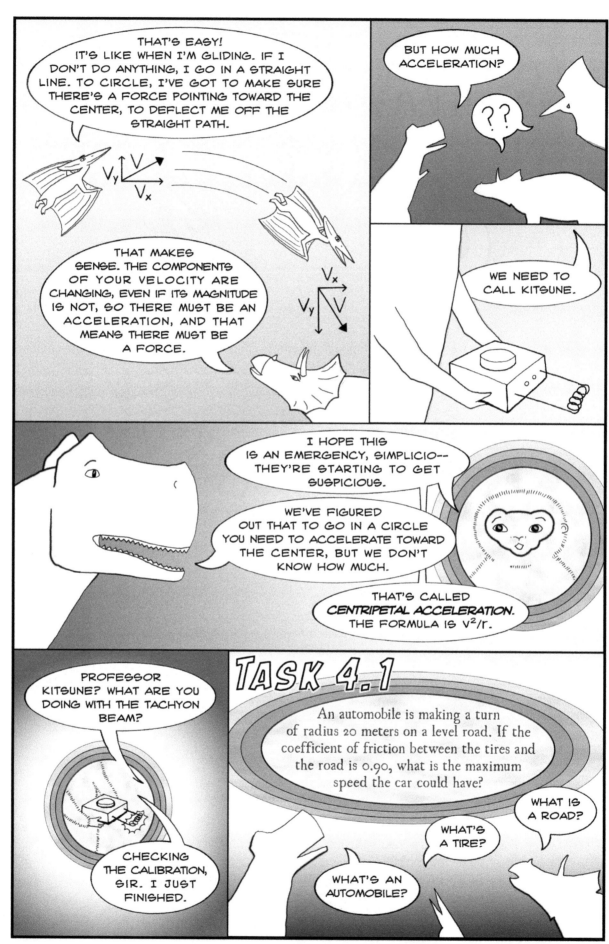

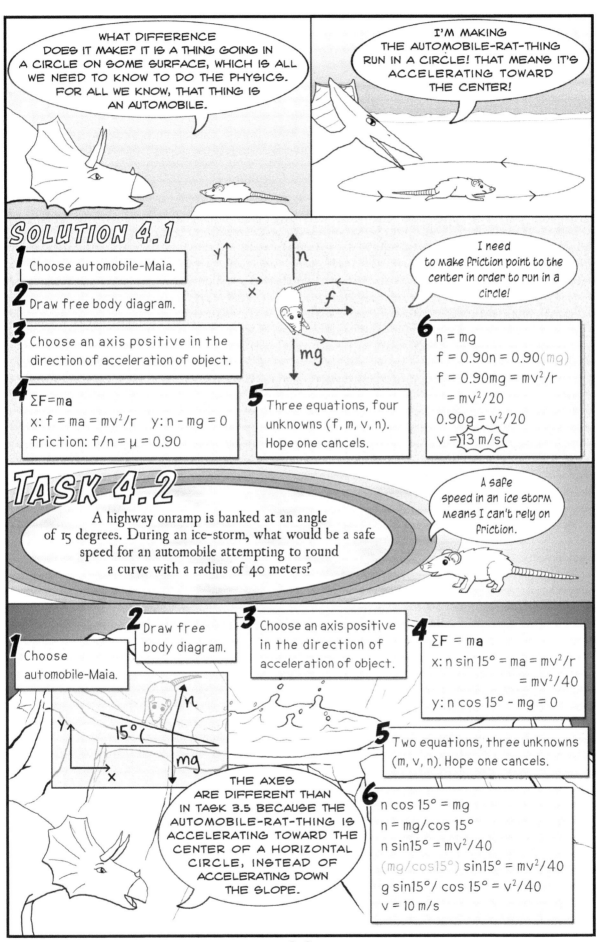

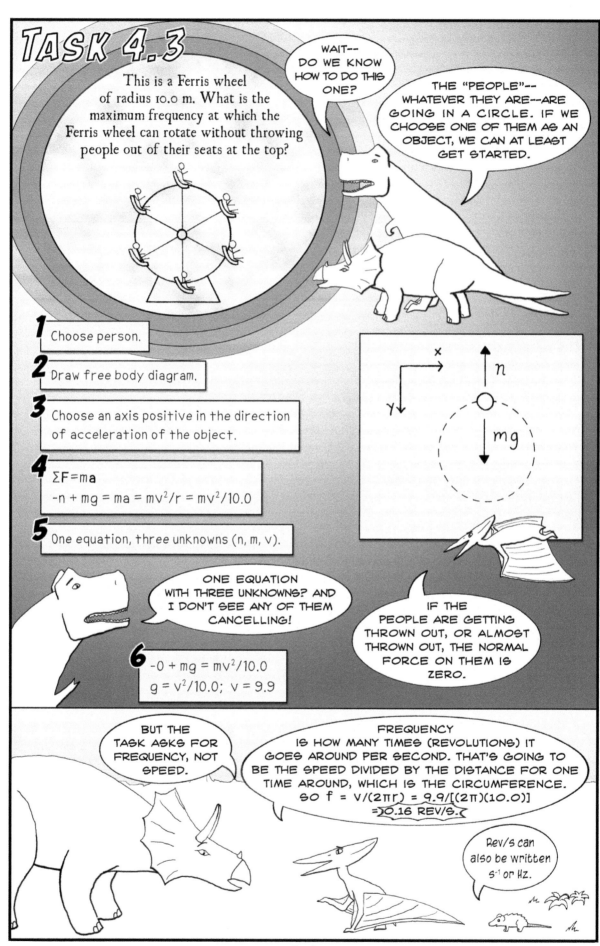

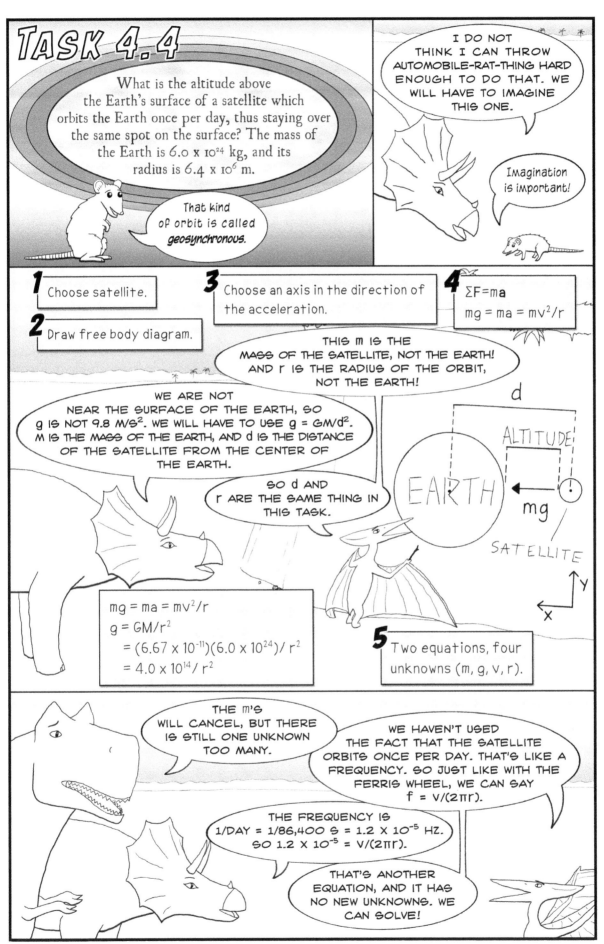

WE'VE GONE FROM THROWING ROCKS ACROSS A CANYON TO LEARNING HOW TO ORBIT THIS EARTH WE LIVE ON AND WHAT GRAVITY IS LIKE ON OTHER PLANETS. BUT WE STILL DON'T KNOW WHY THE PROFESSORS ARE TEACHING ALL OF THIS TO US. HOW DOES IT ALL CONNECT? AND WHERE'S KITSUNE?

EQUATIONS IN THIS CHAPTER

$$g = GM/d^2$$

$$a_{centripetal} = v^2/r$$

$$f = v/(2\pi r)$$

$$f = 1/Period$$

WORDS OF WISDOM

In order to go in a circle, an object needs to accelerate toward the center ("centripetal acceleration")

Proportional reasoning allows us to compare two situations without knowing everything about either situation

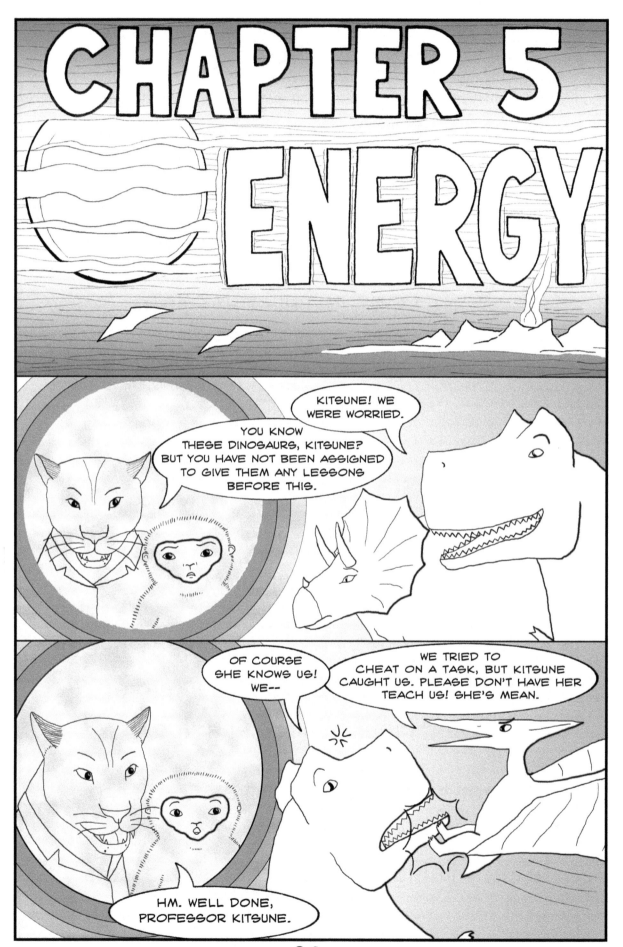

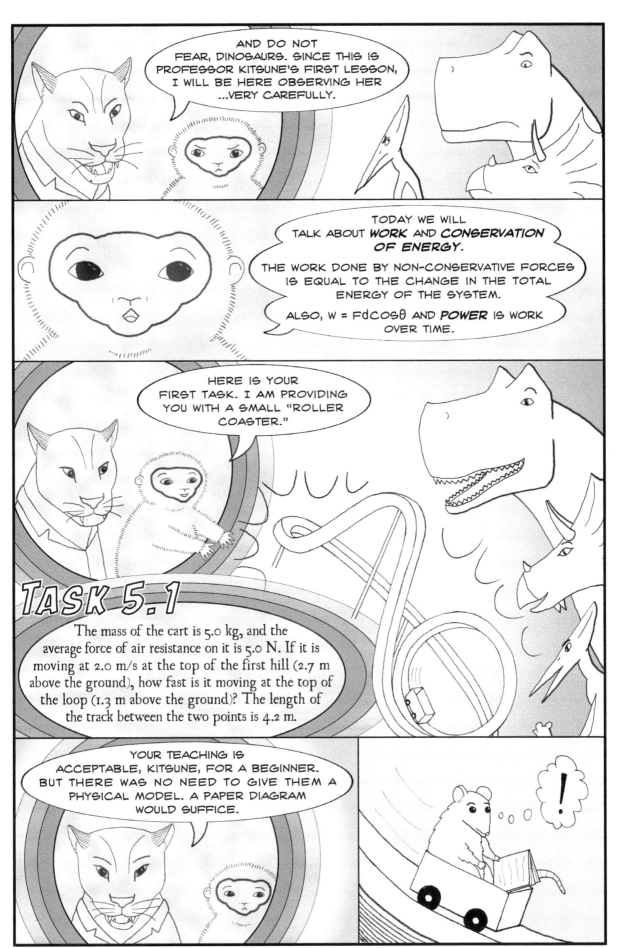

TASK 5.1

The mass of the cart is 5.0 kg, and the average force of air resistance on it is 5.0 N. If it is moving at 2.0 m/s at the top of the first hill (2.7 m above the ground), how fast is it moving at the top of the loop (1.3 m above the ground)? The length of the track between the two points is 4.2 m.

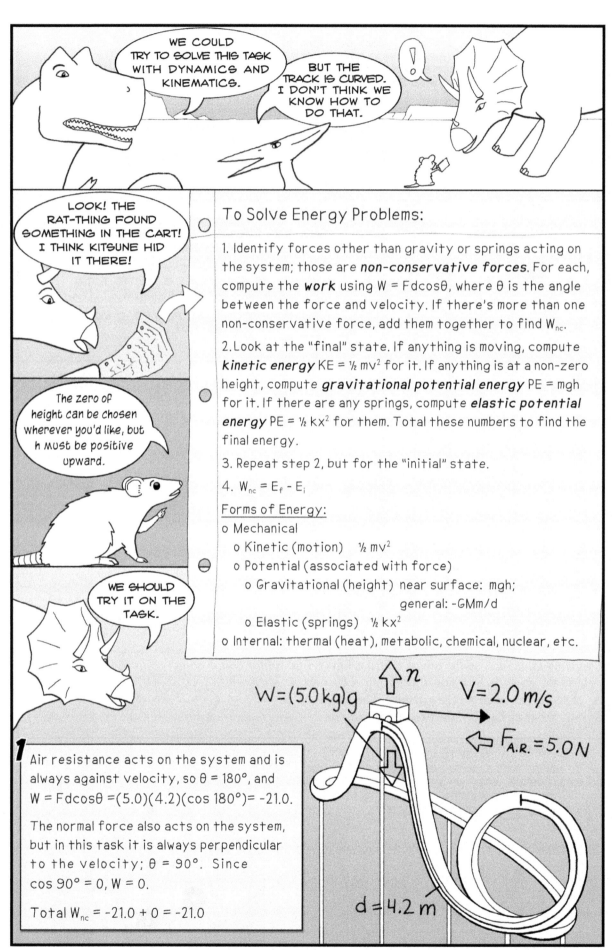

To Solve Energy Problems:

1. Identify forces other than gravity or springs acting on the system; those are **non-conservative forces**. For each, compute the **work** using $W = Fd\cos\theta$, where θ is the angle between the force and velocity. If there's more than one non-conservative force, add them together to find W_{nc}.

2. Look at the "final" state. If anything is moving, compute **kinetic energy** $KE = \frac{1}{2}mv^2$ for it. If anything is at a non-zero height, compute **gravitational potential energy** $PE = mgh$ for it. If there are any springs, compute **elastic potential energy** $PE = \frac{1}{2}kx^2$ for them. Total these numbers to find the final energy.

3. Repeat step 2, but for the "initial" state.

4. $W_{nc} = E_f - E_i$

Forms of Energy:
o Mechanical
  o Kinetic (motion) $\frac{1}{2}mv^2$
  o Potential (associated with force)
    o Gravitational (height) near surface: mgh;
                    general: $-GMm/d$
    o Elastic (springs) $\frac{1}{2}kx^2$
o Internal: thermal (heat), metabolic, chemical, nuclear, etc.

1 Air resistance acts on the system and is always against velocity, so $\theta = 180°$, and $W = Fd\cos\theta = (5.0)(4.2)(\cos 180°) = -21.0$.

The normal force also acts on the system, but in this task it is always perpendicular to the velocity; $\theta = 90°$. Since $\cos 90° = 0$, $W = 0$.

Total $W_{nc} = -21.0 + 0 = -21.0$

2 At top of loop, cart is moving, so
$KE = \frac{1}{2}mv^2 = \frac{1}{2}(5.0)v^2 = 2.5\,v^2$

Cart is at non-zero height, so
$PE = mgh = (5.0)(9.8)(1.3) = 63.7$

No springs.
Final energy $= 2.5\,v^2 + 63.7$

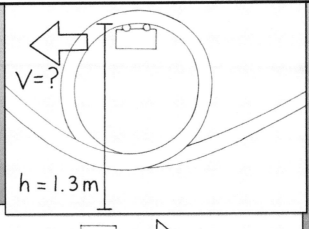

$V = ?$

$h = 1.3\,m$

3 At top of first hill, cart is moving, so
$KE = \frac{1}{2}mv^2 = \frac{1}{2}(5.0)(2.0)^2 = 10.0.$

Cart is at non-zero height, so
$PE = mgh = (5.0)(9.8)(2.7) = 132.3.$

No springs.
Initial energy $= 10.0 + 132.3 = 142.3$

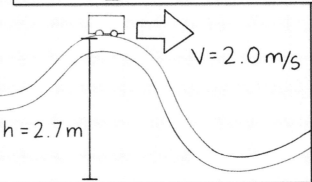

$V = 2.0\,m/s$

$h = 2.7\,m$

4 $W_{nc} = E_f - E_i$
$-21.0 = (2.5v^2 + 63.7) - 142.3$
$57.6 = 2.5\,v^2;$ $v = 4.8\,m/s$

TASK 5.2

A ball is thrown straight up at 13 m/s. How high does it go above its release point? Neglect air resistance.

WHERE'S THE RAT-THING? WE USUALLY USE IT FOR A BALL.

I HAVE NO IDEA. USE A ROCK.

1 No non-conservative forces are acting.

2 At the top of its motion, the ball is instantaneously at rest (not moving). It's at a non-zero height, so $PE = mgh = m(9.8)h = 9.8mh.$

No springs.

$h = ?$ $V = 0$

V

3 Just after release, the ball is moving, so $KE = \frac{1}{2}mv^2 = \frac{1}{2}m(13)^2 = 84.5m$
We'll call the release height zero, and there are no springs.
Initial energy $= 84.5m.$

4 $W_{nc} = E_f - E_i;$ $0 = 9.8mh - 84.5m$
$84.5m = 9.8mh$; m's cancel! $h = 8.6\,m.$

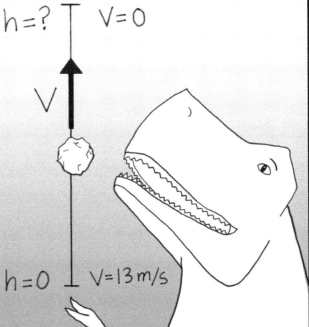

$h = 0$ $V = 13\,m/s$

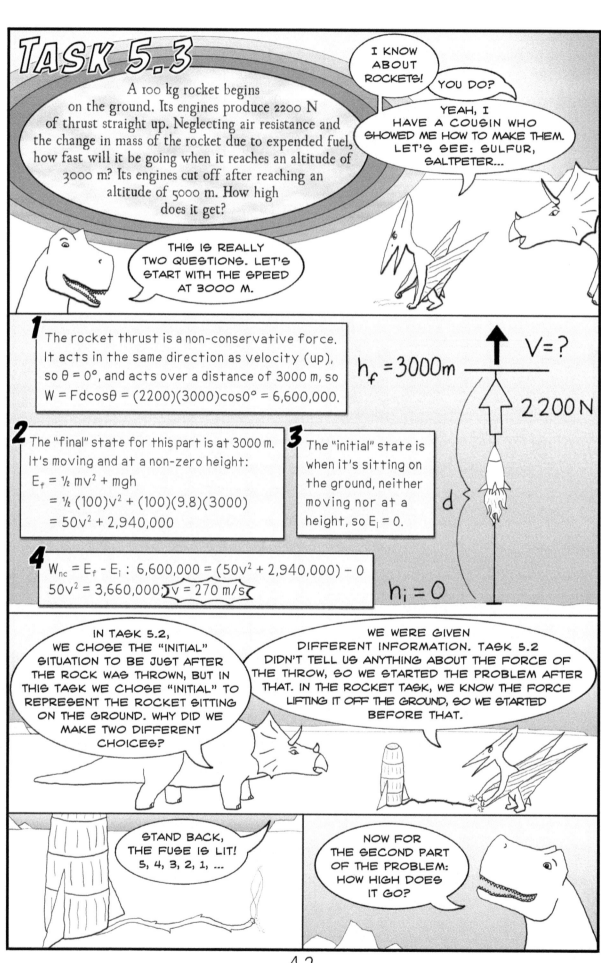

42

SOLUTION 5.3 PART 2

1 This time, the thrust acts over a distance of 5000 m.
$W = Fd\cos\theta = (2200)(5000)\cos 0° = 11{,}000{,}000.$

2 The "final" state is at its highest point. Since it went straight up, its instantaneous speed is zero at that point.

$E_f = mgh = (100)(9.8)h = 980h$

3 The "initial" state hasn't changed, so $E_i = 0$.

4 $W_{nc} = E_f - E_i;$ $11{,}000{,}000 = 980h - 0$
$h = 11{,}200$ m.

WE COULD HAVE SOLVED THIS TASK BY USING KINEMATICS AND DYNAMICS, BUT IT WOULD HAVE TAKEN US MUCH LONGER. WE WOULD HAVE HAD TO FIND THE ACCELERATION WHILE UNDER THRUST USING DYNAMICS, THEN COMPUTED THE VELOCITY AT 5000 M USING KINEMATICS, THEN USED THAT AS AN INITIAL VELOCITY FOR THE PART WHERE THERE WAS NO THRUST. ENERGY LET US DO THIS TASK MUCH MORE QUICKLY!

In reality, air resistance would be acting opposite the velocity and thus doing negative work, like in the roller coaster task (Task 5.1). And the rocket would also be throwing part of its mass backward, because that's what actually provides the thrust. Textbooks often tell students to neglect air resistance and the change in rocket mass for tasks like this in order to keep the task from being too hard, even though neither assumption is very realistic.

$h_f = ?$

$h = 5000$ m

2200 N

d

$(100 kg)g$

$h_i = 0$

TASK 5.4

A 1.50 kg egg is dropped from a height of 2.00 m above sand. The egg can withstand a maximum force of 410.0 N without cracking. To the nearest mm, what is the minimum distance it has to penetrate into the sand to avoid cracking? Assume the sand exerts a constant force on the egg.

1 The upward force of the sand on the egg is a non-conservative force. We want the case where it's at its maximum, 410.0 N. The force acts opposite the velocity, so $\theta = 180°$.

$W = Fd\cos\theta = (410.0)d\cos 180° = -410.0d$

2 If we choose the surface of the sand as $h = 0$, then the egg ends at a non-zero height, but at rest.

$E_f = mgh = (1.50)(9.8)(-d) = -14.7d.$

$h = 2$ m

$h = 0$
$h_f = ?$
d
410.0 N

3 Simplicio drops the egg, so the initial speed is zero, but it's at a height of 2.00 m.

$E_i = mgh = (1.50)(9.8)(2.00)$
$= 29.4$

4 $W_{nc} = E_f - E_i$
$-410.0d = -14.7d - 29.4$
$-395.3d = -29.4$
$d = 0.074 \text{ m} = 74 \text{ mm}$

E_i

h

$E_f \quad d \quad O$

TASK 5.5

Simplicio (mass 6000 kg) jumps on a trampoline from a height of 2.30 m above its surface. If he sinks a maximum of 0.40 m into the trampoline, what is its effective spring constant?

1 We're told to treat the trampoline like a spring, so the force it exerts is conservative. No non-conservative forces act in this task.

2 "Final" will be when the trampoline is maximally compressed. If we choose h = 0 as the initial surface of the trampoline, there is a non-zero height of -0.40 m and a spring that is compressed by 0.40 m.

$E_f = mgh + \frac{1}{2}kx^2$
$= (6000)(9.8)(-0.40) + \frac{1}{2}k(0.40)^2$
$= -23,500 + 0.080k$

3 "Initial" is when Simplicio jumps from 2.30 m. Since we're not told otherwise, we'll assume his initial velocity is negligible. The trampoline is not yet compressed, so the only kind of energy is gravitational potential energy.

$E_i = mgh = (6000)(9.8)(2.30) = 135,000$

4 $W_{nc} = E_f - E_i; \quad 0 = -23,500 + 0.080k - 135,000$
$158,500 = 0.080k; \quad k = 2,000,000 \text{ N/m}$

$h_i = 2.3 \text{ m}$

E_i

$h = 0$

$h_f = -0.4 \text{m}$

E_f

$W = 6000 \text{kg} \cdot g$

That's a big spring constant, but Simplicio is a big dinosaur. I am about 2.0 kg, so if I jumped from 2.30 m onto a trampoline and it compressed by 0.40 m, the spring constant would work out to a much smaller 660 N/m. Some of the textbooks, though, have examples with spring constants less than 10 N/m...those are very weak springs!

TASK 5.6

What is the speed of the 5.0 kg block just before it hits the ground? The system is released from rest in the position shown.

1 Tension is a non-conservative force, but it acts within this system, not on it, since the objects at both ends of the rope are included in our system. The force that holds the pulley up does no work because the pulley does not move up or down. So there are no non-conservative forces doing work on this system.

2 "Final" will be just before the 5.0 kg block hits the ground. At that point, the 3.0 kg block is moving, the 5.0 kg block is moving, and the 3.0 kg block is at a non-zero height. So $E_f = \frac{1}{2} m_3 v_3{}^2 + \frac{1}{2} m_5 v_5{}^2 + m_3 g h_3$.

The two blocks have the same speed (though in opposite directions) because they're connected by a rope, so $v_3 = v_5 = v$. If the 5.0 kg block goes 20 cm down, the 3.0 kg block must go 20 cm up, so $h_3 = 40$ cm $= 0.40$ m. Thus, $E_f = \frac{1}{2}(3.0)v^2 + \frac{1}{2}(5.0)v^2 + (3.0)(9.8)(0.40) = 4.0v^2 + 11.8$.

20cm

20cm $V_1 \uparrow$ $V_2 \downarrow$

3 kg   5 kg

3 "Initial" is when the system is released. Nothing is moving yet, and both blocks are 20 cm $= 0.20$ m above the ground.
$E_i = m_3 g h_3 + m_5 g h_5 = (3.0)(9.8)(0.20) + (5.0)(9.8)(0.20) = 15.7$.

4 $W_{nc} = E_f - E_i$
$0 = (4.0v^2 + 11.8) - 15.7$
$3.9 = 4.0v^2$; $v = 0.99$ m/s

I LIKE HOW IN ENERGY PROBLEMS WE CAN JUST ADD EVERYTHING TOGETHER. NO VECTORS, NO SEPARATE FREE-BODY DIAGRAMS, NO WORRYING ABOUT DIFFERENT COORDINATE SYSTEMS!

TASK 5.7

How much average power is needed to raise an elevator and its passengers (total mass = 800.0 kg) a distance of 6.0 meters in 4.0 seconds?

KITSUNE DIDN'T LEAVE US INSTRUCTIONS ON HOW TO SOLVE POWER PROBLEMS!

BUT WHAT'S AN ELEVATOR?

KITSUNE DID SAY POWER IS WORK OVER TIME. I THINK WE SHOULD START OUT USING OUR WORK-ENERGY PROCEDURE, AND SEE IF WE CAN USE THAT DEFINITION.

IT'S LIKE "AUTOMOBILE." WE DON'T HAVE TO KNOW WHAT IT IS TO SOLVE THE TASK.

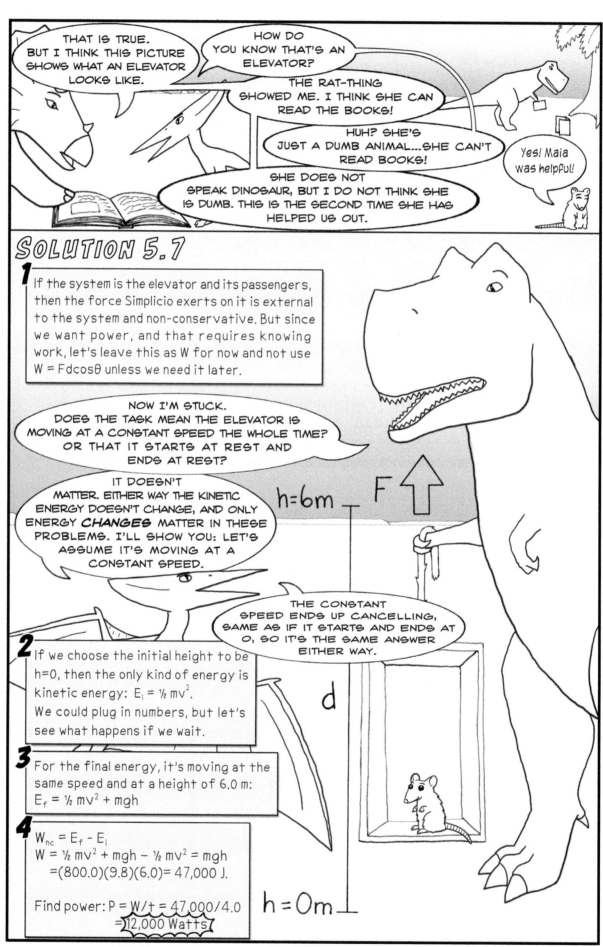

THAT IS TRUE. BUT I THINK THIS PICTURE SHOWS WHAT AN ELEVATOR LOOKS LIKE.

HOW DO YOU KNOW THAT'S AN ELEVATOR?

THE RAT-THING SHOWED ME. I THINK SHE CAN READ THE BOOKS!

HUH? SHE'S JUST A DUMB ANIMAL...SHE CAN'T READ BOOKS!

Yes! Maia was helpful!

SHE DOES NOT SPEAK DINOSAUR, BUT I DO NOT THINK SHE IS DUMB. THIS IS THE SECOND TIME SHE HAS HELPED US OUT.

SOLUTION 5.7

1 If the system is the elevator and its passengers, then the force Simplicio exerts on it is external to the system and non-conservative. But since we want power, and that requires knowing work, let's leave this as W for now and not use $W = Fd\cos\theta$ unless we need it later.

NOW I'M STUCK. DOES THE TASK MEAN THE ELEVATOR IS MOVING AT A CONSTANT SPEED THE WHOLE TIME? OR THAT IT STARTS AT REST AND ENDS AT REST?

IT DOESN'T MATTER. EITHER WAY THE KINETIC ENERGY DOESN'T CHANGE, AND ONLY ENERGY *CHANGES* MATTER IN THESE PROBLEMS. I'LL SHOW YOU: LET'S ASSUME IT'S MOVING AT A CONSTANT SPEED.

THE CONSTANT SPEED ENDS UP CANCELLING, SAME AS IF IT STARTS AND ENDS AT 0, SO IT'S THE SAME ANSWER EITHER WAY.

$h = 6m$

F

d

2 If we choose the initial height to be h=0, then the only kind of energy is kinetic energy: $E_i = \frac{1}{2}mv^2$.
We could plug in numbers, but let's see what happens if we wait.

3 For the final energy, it's moving at the same speed and at a height of 6.0 m:
$E_f = \frac{1}{2}mv^2 + mgh$

4 $W_{nc} = E_f - E_i$
$W = \frac{1}{2}mv^2 + mgh - \frac{1}{2}mv^2 = mgh$
$= (800.0)(9.8)(6.0) = 47,000$ J.

Find power: $P = W/t = 47,000/4.0$
$= 12,000$ Watts

$h = 0m$

46

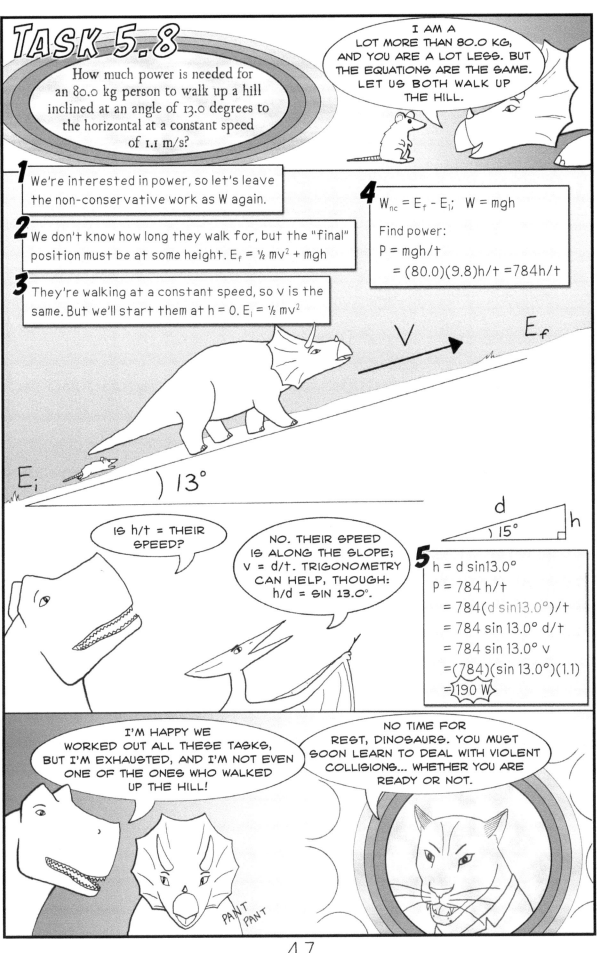

EQUATIONS IN THIS CHAPTER

$$W_{nc} = E_f - E_i$$

$$W = Fd\cos\theta$$

$$KE = 1/2\ mv^2$$

$$PE_{grav} = mgh$$

$$PE_{elastic} = 1/2\ kx^2$$

$$P = W/t$$

WORDS OF WISDOM

Work is change in energy

Non-conservative forces do not have potential energies associated with them

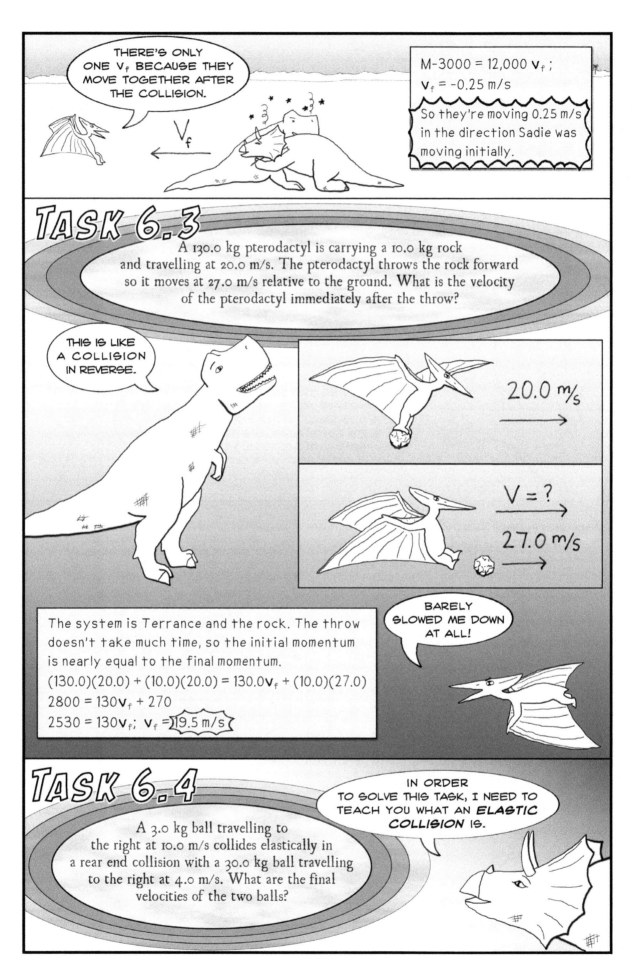

THERE'S ONLY ONE V_f BECAUSE THEY MOVE TOGETHER AFTER THE COLLISION.

V_f

M–3000 = 12,000 v_f;
v_f = –0.25 m/s

So they're moving 0.25 m/s in the direction Sadie was moving initially.

TASK 6.3

A 130.0 kg pterodactyl is carrying a 10.0 kg rock and travelling at 20.0 m/s. The pterodactyl throws the rock forward so it moves at 27.0 m/s relative to the ground. What is the velocity of the pterodactyl immediately after the throw?

THIS IS LIKE A COLLISION IN REVERSE.

20.0 m/s →

V = ? →
27.0 m/s →

The system is Terrance and the rock. The throw doesn't take much time, so the initial momentum is nearly equal to the final momentum.
(130.0)(20.0) + (10.0)(20.0) = 130.0v_f + (10.0)(27.0)
2800 = 130v_f + 270
2530 = 130v_f; v_f = 19.5 m/s

BARELY SLOWED ME DOWN AT ALL!

TASK 6.4

A 3.0 kg ball travelling to the right at 10.0 m/s collides elastically in a rear end collision with a 30.0 kg ball travelling to the right at 4.0 m/s. What are the final velocities of the two balls?

IN ORDER TO SOLVE THIS TASK, I NEED TO TEACH YOU WHAT AN *ELASTIC COLLISION* IS.

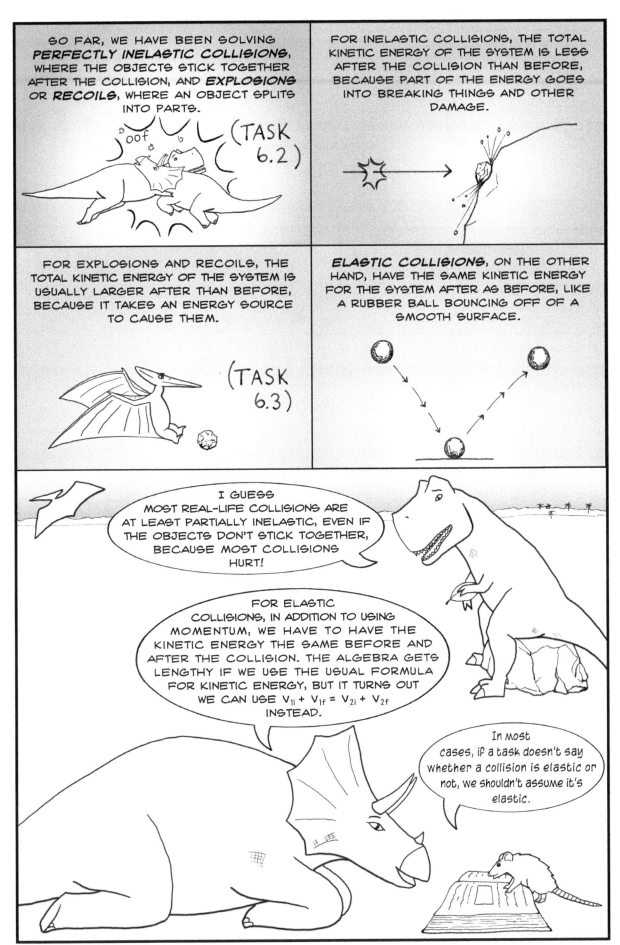

SO FAR, WE HAVE BEEN SOLVING **PERFECTLY INELASTIC COLLISIONS**, WHERE THE OBJECTS STICK TOGETHER AFTER THE COLLISION, AND **EXPLOSIONS** OR **RECOILS**, WHERE AN OBJECT SPLITS INTO PARTS.

oof

(TASK 6.2)

FOR INELASTIC COLLISIONS, THE TOTAL KINETIC ENERGY OF THE SYSTEM IS LESS AFTER THE COLLISION THAN BEFORE, BECAUSE PART OF THE ENERGY GOES INTO BREAKING THINGS AND OTHER DAMAGE.

FOR EXPLOSIONS AND RECOILS, THE TOTAL KINETIC ENERGY OF THE SYSTEM IS USUALLY LARGER AFTER THAN BEFORE, BECAUSE IT TAKES AN ENERGY SOURCE TO CAUSE THEM.

(TASK 6.3)

ELASTIC COLLISIONS, ON THE OTHER HAND, HAVE THE SAME KINETIC ENERGY FOR THE SYSTEM AFTER AS BEFORE, LIKE A RUBBER BALL BOUNCING OFF OF A SMOOTH SURFACE.

I GUESS MOST REAL-LIFE COLLISIONS ARE AT LEAST PARTIALLY INELASTIC, EVEN IF THE OBJECTS DON'T STICK TOGETHER, BECAUSE MOST COLLISIONS HURT!

FOR ELASTIC COLLISIONS, IN ADDITION TO USING MOMENTUM, WE HAVE TO HAVE THE KINETIC ENERGY THE SAME BEFORE AND AFTER THE COLLISION. THE ALGEBRA GETS LENGTHY IF WE USE THE USUAL FORMULA FOR KINETIC ENERGY, BUT IT TURNS OUT WE CAN USE $v_{1i} + v_{1f} = v_{2i} + v_{2f}$ INSTEAD.

In most cases, if a task doesn't say whether a collision is elastic or not, we shouldn't assume it's elastic.

SOLUTION 6.4

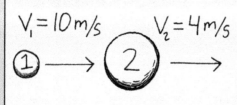  

If we make to the right positive, then momentum tells us:

$(3.0)(10.0) + (30.0)(4.0)$
$\qquad = (3.0)\, v_{1f} + (30.0)\, v_{2f}$

Since the collision is elastic, $10.0 + v_{1f} = 4.0 + v_{2f}$

We have two equations and two unknowns, so we can solve.

From the second equation: $v_{2f} = 6.0 + v_{1f}$
Substituting that in the first equation:
$150.0 = (3.0)\, v_{1f} + (30.0)\,(6.0 + v_{1f})$
$150.0 = (3.0)\, v_{1f} + 180.0 + 30.0\, v_{1f}$
$-30.0 = 33.0\, v_{1f}$
$v_{1f} = -0.91\ m/s$
$v_{2f} = 6.0 + (-0.91) = 5.1\ m/s$
The negative means the lighter ball bounces back toward the left.

THAT'S IT? I GUESS COLLISIONS AREN'T SO BAD!

EQUATIONS IN THIS CHAPTER

$$J_{external} = \Delta p$$

$$J = Ft$$

$$p = mv$$

Elastic collision: $v_{1i} + v_{1f} = v_{2i} + v_{2f}$

WORDS OF WISDOM

Mechanical (kinetic) energy is conserved for elastic collisions, but is not for inelastic

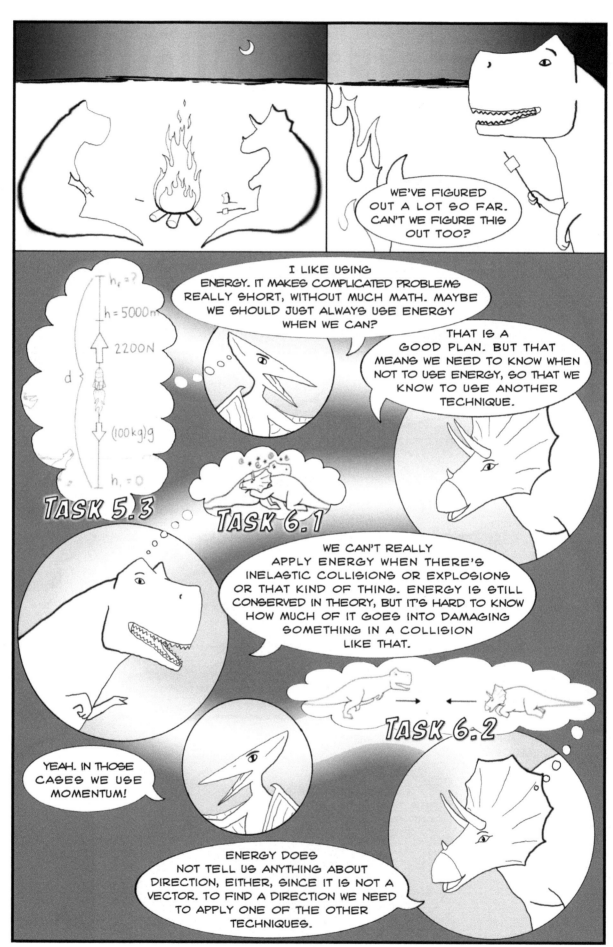

56

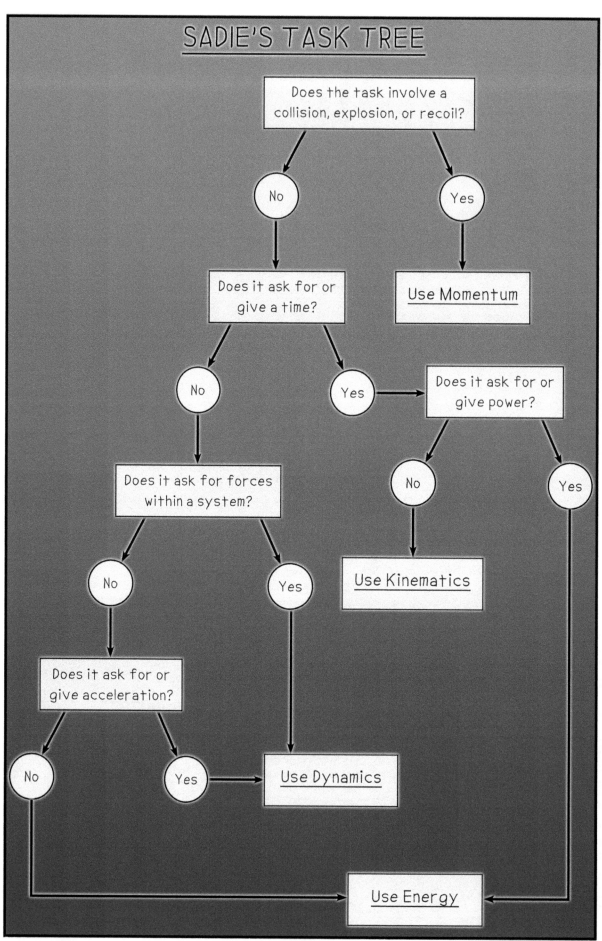

CHAPTER 7
ROTATIONAL PHYSICS

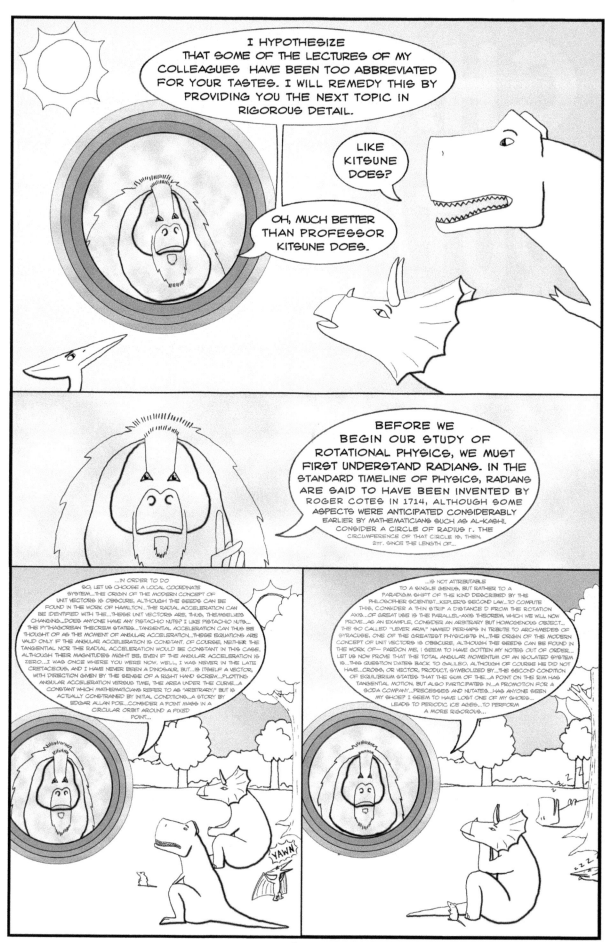

60

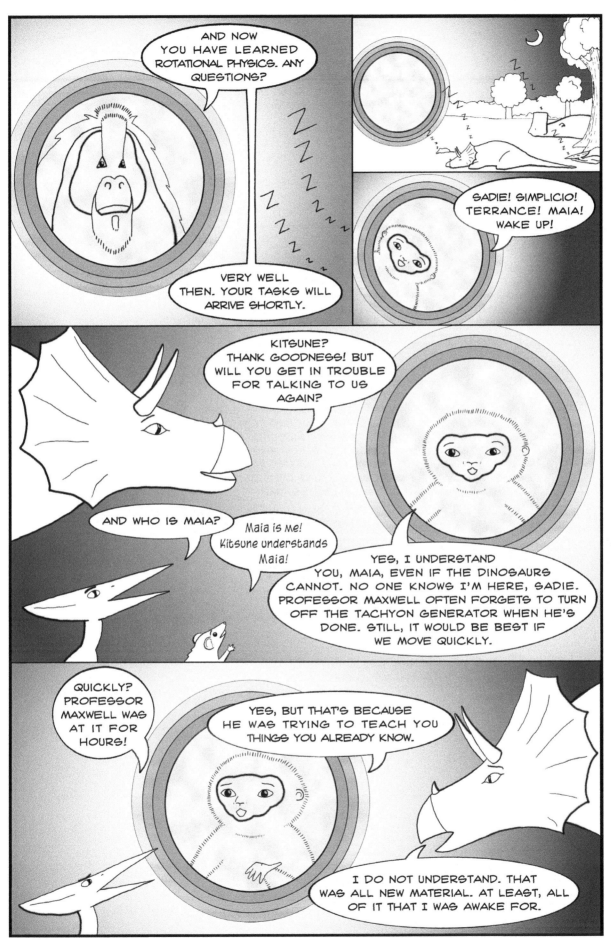

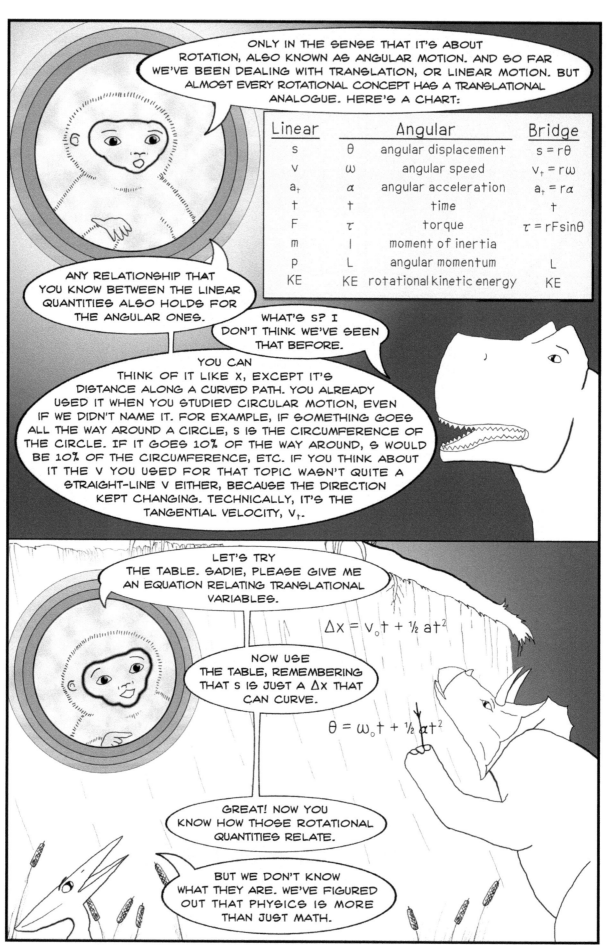

ONLY IN THE SENSE THAT IT'S ABOUT ROTATION, ALSO KNOWN AS ANGULAR MOTION. AND SO FAR WE'VE BEEN DEALING WITH TRANSLATION, OR LINEAR MOTION. BUT ALMOST EVERY ROTATIONAL CONCEPT HAS A TRANSLATIONAL ANALOGUE. HERE'S A CHART:

Linear	Angular		Bridge
s	θ	angular displacement	$s = r\theta$
v	ω	angular speed	$v_t = r\omega$
a_t	α	angular acceleration	$a_t = r\alpha$
t	t	time	t
F	τ	torque	$\tau = rF\sin\theta$
m	I	moment of inertia	
p	L	angular momentum	L
KE	KE	rotational kinetic energy	KE

ANY RELATIONSHIP THAT YOU KNOW BETWEEN THE LINEAR QUANTITIES ALSO HOLDS FOR THE ANGULAR ONES.

WHAT'S S? I DON'T THINK WE'VE SEEN THAT BEFORE.

YOU CAN THINK OF IT LIKE X, EXCEPT IT'S DISTANCE ALONG A CURVED PATH. YOU ALREADY USED IT WHEN YOU STUDIED CIRCULAR MOTION, EVEN IF WE DIDN'T NAME IT. FOR EXAMPLE, IF SOMETHING GOES ALL THE WAY AROUND A CIRCLE, S IS THE CIRCUMFERENCE OF THE CIRCLE. IF IT GOES 10% OF THE WAY AROUND, S WOULD BE 10% OF THE CIRCUMFERENCE, ETC. IF YOU THINK ABOUT IT THE V YOU USED FOR THAT TOPIC WASN'T QUITE A STRAIGHT-LINE V EITHER, BECAUSE THE DIRECTION KEPT CHANGING. TECHNICALLY, IT'S THE TANGENTIAL VELOCITY, v_t.

LET'S TRY THE TABLE. SADIE, PLEASE GIVE ME AN EQUATION RELATING TRANSLATIONAL VARIABLES.

$$\Delta x = v_0 t + \tfrac{1}{2} a t^2$$

NOW USE THE TABLE, REMEMBERING THAT S IS JUST A Δx THAT CAN CURVE.

$$\theta = \omega_0 t + \tfrac{1}{2} \alpha t^2$$

GREAT! NOW YOU KNOW HOW THOSE ROTATIONAL QUANTITIES RELATE.

BUT WE DON'T KNOW WHAT THEY ARE. WE'VE FIGURED OUT THAT PHYSICS IS MORE THAN JUST MATH.

TASK 7.1

A phonograph record is designed to play at 33.3 rev/min. If it takes 4.0 s to get up to speed when it is turned on, what is its angular acceleration? How many revs does it make in that time?

WHAT IS A "PHONOGRAPH RECORD"? I BET IT'S SOME SUPER-ADVANCED PIECE OF TECHNOLOGY FROM THE DISTANT FUTURE!

IT DOESN'T MATTER--WE'VE GOT THIS. IT SOUNDS LIKE THE THING STARTS AT REST. AFTER THAT, A LOT OF IT'S JUST UNITS. WE NEED ANGLES IN RADIANS, AND EVERYTHING IN SI.

$33.3 \, \text{rev/min} * 1 \, \text{min}/60 \, \text{s} * 2\pi \, \text{rad/rev} = 3.49 \, \text{rad/s}$

Analogy to $v = v_o + at$

$\omega = \omega_o + \alpha t = 3.49 = 0 + \alpha(4.0)$

$\alpha = 0.87 \, \text{rad/s}^2$

Analogy to $\Delta x = v_o t + \frac{1}{2} at^2$

$\theta = \omega_o t + \frac{1}{2} \alpha t^2$

$\theta = 0 \, (4.0) + \frac{1}{2} \, (0.87)(4.0)^2 = 7.0 \, \text{rads}$

TASK 7.2

A 0.70 m radius wheel starts from rest and accelerates at 3.0 rad/s² for 1.0 s. How many m does it travel? What is the tangential acceleration of a point on the outside of the wheel at the end of the 3.0 s? What is the radial acceleration at the same instant?

I KNOW, SIMPLICIO, "WHAT'S A WHEEL?" HERE...I'VE JUST MADE THIS THING THAT SPINS. LET'S SAY I'VE INVENTED A WHEEL AND MOVE ON.

$\omega = \omega_o + \alpha t$

$\omega = 0 + (3.0)(1.0) = 3.0$

$\theta = \omega_o t + \frac{1}{2} \alpha t^2$

$\theta = 0(1.0) + \frac{1}{2} \, (3.0)(1.0)^2 = 1.5$

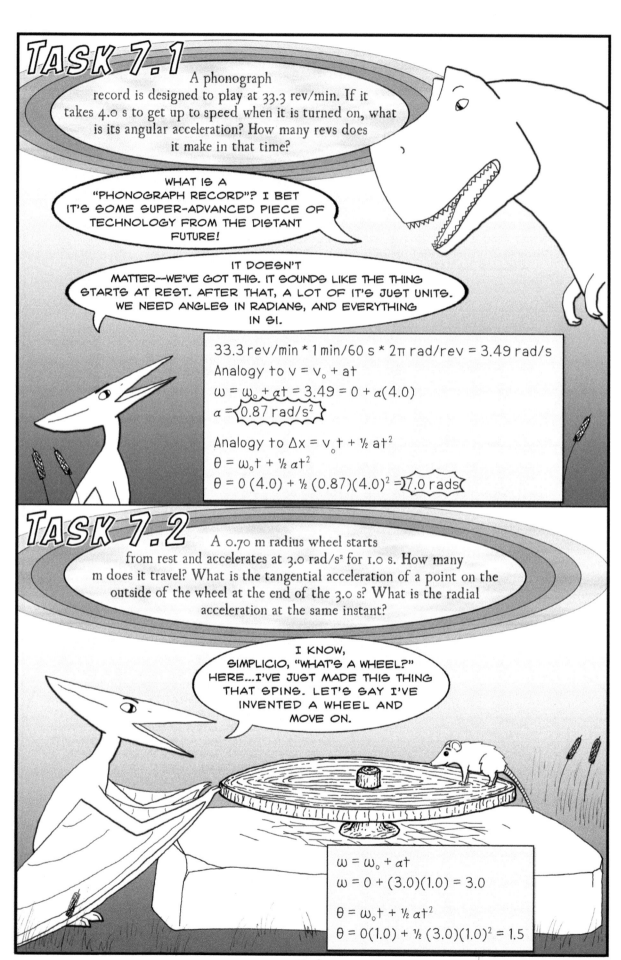

64

But we're asked for distance in m, so we need a bridge equation:

$$s = r\theta = 0.70(1.5) = \boxed{1.0\text{ m}}$$

Likewise, a bridge equation gets us a_t, the tangential acceleration:

$$a_t = r\alpha = 0.70(3.0) = \boxed{2.1\text{ m/s}^2}$$

Radial acceleration means along the center and centripetal acceleration means toward the center, so they're different ways of specifying the same magnitude:

$$a_c = v^2/r$$

We don't have v, but we can use another bridge equation:

$$v_t = r\omega = 0.70(3.0) = 2.1$$
$$a_c = 2.1^2/0.70 = \boxed{6.3\text{ m/s}^2}$$

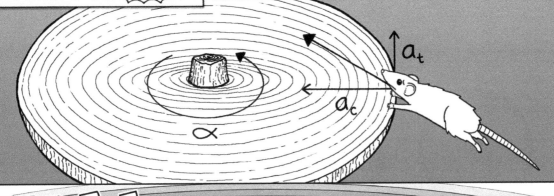

TASK 7.3

A 50.0 kg block and a 30.0 kg block are attached to opposite ends up a rope that is looped over a frictionless pulley. The pulley is a uniform cylinder with a mass of 100.0 kg and a radius of 0.75 m. What is the acceleration of the blocks and the tension in each side of the rope?

HAVEN'T WE DONE THIS KIND OF TASK BEFORE?

THIS ONE IS DIFFERENT, BECAUSE WE ARE GIVEN THE MASS OF THE PULLEY. IN THE DYNAMICS TASK, TERRANCE NEGLECTED THE MASS OF THE PULLEYS AS TOO SMALL TO MATTER MUCH. WE DID IT AGAIN WITH THE ENERGETICS TASK.

I WISH I COULD NEGLECT THE MASS OF THIS PULLEY--IT'S HEAVY!

Textbooks expect us to neglect the mass of pulleys unless their mass is given or asked for.

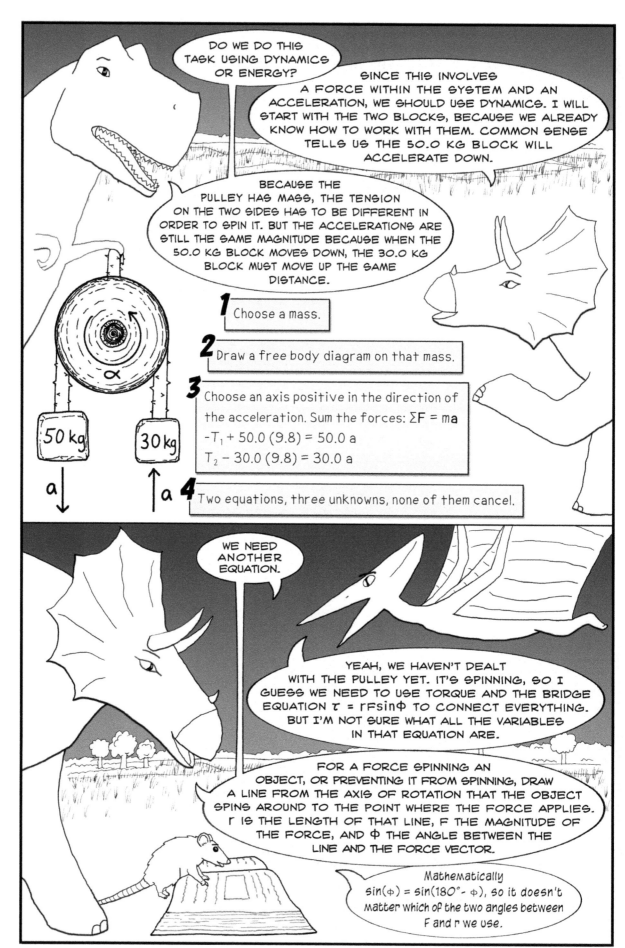

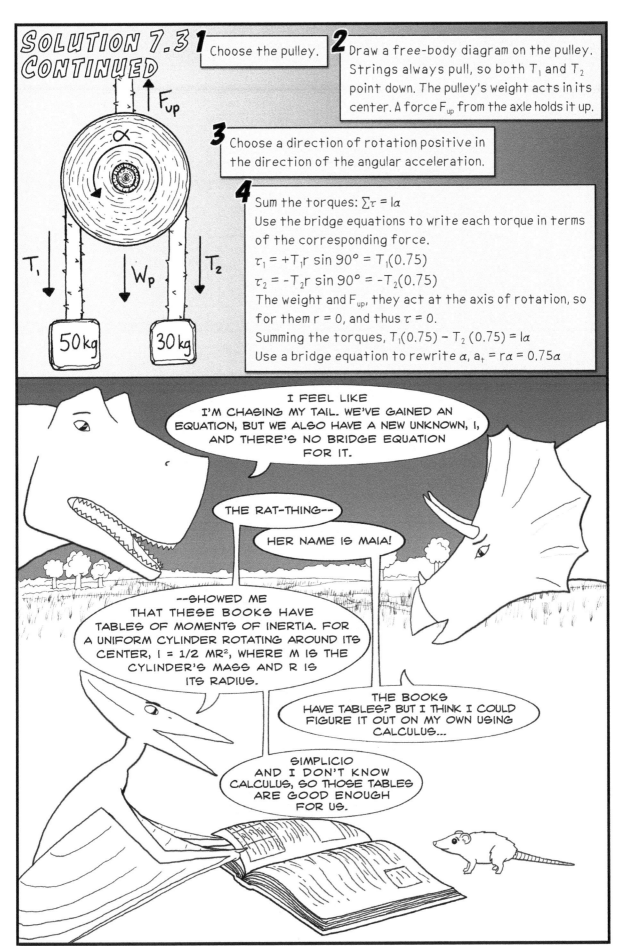

SOLUTION 7.31 CONTINUED

1 Choose the pulley.

2 Draw a free-body diagram on the pulley. Strings always pull, so both T_1 and T_2 point down. The pulley's weight acts in its center. A force F_{up} from the axle holds it up.

3 Choose a direction of rotation positive in the direction of the angular acceleration.

4 Sum the torques: $\sum \tau = I\alpha$
Use the bridge equations to write each torque in terms of the corresponding force.
$\tau_1 = +T_1 r \sin 90° = T_1(0.75)$
$\tau_2 = -T_2 r \sin 90° = -T_2(0.75)$
The weight and F_{up}, they act at the axis of rotation, so for them $r = 0$, and thus $\tau = 0$.
Summing the torques, $T_1(0.75) - T_2(0.75) = I\alpha$
Use a bridge equation to rewrite α, $a_t = r\alpha = 0.75\alpha$

F_{up}

α

T_1 W_p T_2

50 kg   30 kg

I FEEL LIKE I'M CHASING MY TAIL. WE'VE GAINED AN EQUATION, BUT WE ALSO HAVE A NEW UNKNOWN, I, AND THERE'S NO BRIDGE EQUATION FOR IT.

THE RAT-THING--

HER NAME IS MAIA!

--SHOWED ME THAT THESE BOOKS HAVE TABLES OF MOMENTS OF INERTIA. FOR A UNIFORM CYLINDER ROTATING AROUND ITS CENTER, I = 1/2 MR², WHERE M IS THE CYLINDER'S MASS AND R IS ITS RADIUS.

THE BOOKS HAVE TABLES? BUT I THINK I COULD FIGURE IT OUT ON MY OWN USING CALCULUS...

SIMPLICIO AND I DON'T KNOW CALCULUS, SO THOSE TABLES ARE GOOD ENOUGH FOR US.

WE'VE GOT A LOT OF EQUATIONS! I'LL NUMBER THEM.

5

1: $-T_1 + 50.0\,(9.8) = 50.0\,a$ 2: $T_2 - 30.0\,(9.8) = 30.0\,a$

3: $T_1(0.75) - T_2(0.75) = I\alpha$ 4: $a = a_t = 0.75\alpha$

5: $I = \frac{1}{2}MR^2 = \frac{1}{2}(100.0)(0.75)^2 = 28.1$

Unknowns: T_1, T_2, a, I, α

5 equations and 5 unknowns, so we can solve!

6

Eq. 1: $T_1 = 490 - 50.0\,a$ Eq. 2: $T_2 = 30.0\,a + 294$

Eq. 4: $\alpha = a/0.75$

Eq. 3: $(490 - 50.0\,a)(0.75) - (30.0\,a + 294)(0.75) = (28.1)(a/0.75)$

$368 - 37.5\,a - 22.5\,a - 220 = 37.5\,a$

$148 = 97.5a$; $a = 1.52\ \text{m/s}^2$

$T_1 = 490 - 50.0(1.52) = 414\ \text{N}$; $T_2 = 30.0(1.52) + 294 = 340\ \text{N}$

TASK 7.4

A solid sphere is rolled down a 3.50 m long plane inclined at an angle of 30.0° to the horizontal. How fast is it travelling when it reaches the bottom?

WE SHOULD BE ABLE TO DO THIS ONE WITH ENERGY!

I AM A LITTLE CONFUSED ABOUT ROTATIONAL ENERGY. THE BRIDGE EQUATION TABLE LISTS ROTATIONAL KE, BUT DOES NOT GIVE US AN EQUATION.

MAYBE WE CAN DO IT USING THE ANALOGY METHOD INSTEAD. TRANSLATIONAL KE = 1/2 mv², SO ROTATIONAL KE = 1/2 Iω².

THAT MAKES SENSE. THOSE ARE TWO DIFFERENT KINDS OF ENERGY. WHEN SOMETHING TRANSLATES, IT HAS 1/2 mv², WHEN IT SPINS, IT HAS 1/2 Iω².

IN THIS TASK, THE SPHERE ROLLS. ROLLING INVOLVES BOTH TRANSLATION AND SPINNING, SO IT WILL HAVE BOTH KINDS OF ENERGY.

1 The normal force is non-conservative, but acts perpendicular to the velocity, so there is no work done by non-conservative forces.

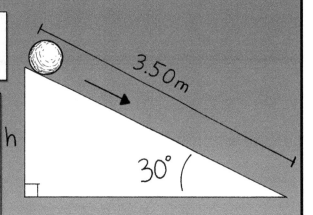

2 The final state is at the bottom of the incline.
The sphere is moving: $KE = \frac{1}{2} mv^2$.
The sphere is spinning: $KE = \frac{1}{2} I\omega^2$.
If we choose the bottom of the ramp as height 0, $PE = 0$.
Final energy $= \frac{1}{2} mv^2 + \frac{1}{2} I\omega^2$

3 Initially, at the top of the incline, the sphere is not moving or spinning, but it is at a non-zero height. Trigonometry gives the height as $3.50 \sin(30.0°) = 1.75$.
Initial energy $= PE = mgh = m(9.8)(1.75)$

4 $W_{nc} = E_f - E_i$
$0 = \frac{1}{2} mv^2 + \frac{1}{2} I\omega^2 - 17.2m$
For a sphere, $I = 2/5 MR^2$
$0 = \frac{1}{2} mv^2 + \frac{1}{2} (2/5 MR^2) \omega^2 - 17.2m$
m's cancel: $17.2 = \frac{1}{2} v^2 + 1/5 r^2\omega^2$

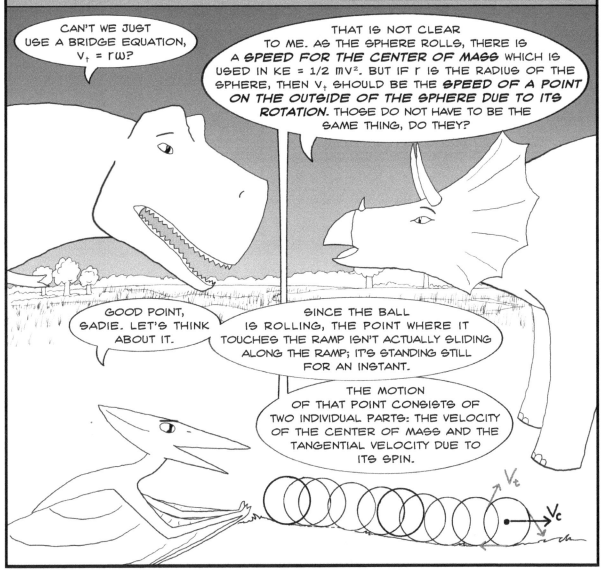

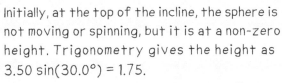

TASK 7.5

A 120.0 cm diameter uniform disc with a mass of 30.0 kg is dropped on to a 200.0 cm diameter uniform disc with a mass of 100.0 kg. The smaller disc was originally not rotating, but the larger one was rotating at 2.0 rev/s. If the two discs stick together after the collision, what is their mutual angular speed after the collision?

THIS ONE'S EASY! IT'S A COLLISION, SO WE'LL USE CONSERVATION OF MOMENTUM.

Analogy to $p = mv$: $L = I\omega$

The system is both discs. The angular momentum before the collision is the same as after:
$$I_{1i}\omega_{1i} + I_{2i}\omega_{2i} = I_{1f}\omega_{1f} + I_{2f}\omega_{2f}$$

For a uniform disc spinning around its center, $I = \frac{1}{2} MR^2$.
$$I_1 = \frac{1}{2}(30.0)(1.200)^2 = 21.6$$
$$I_2 = \frac{1}{2}(100.0)(2.000)^2 = 200.0$$

Convert 2.0 rev/s to rad/s:
$$2.0 \text{ rev/s} * 2\pi \text{ rad/rev} = 12.6 \text{ rad/s}$$

$$(21.6)(12.6) + (200.0)(0) = (21.6)\omega_f + (200.0)\omega_f$$
$$272 = 221.6\omega_f$$
$$\omega_f = 1.23 \text{ rad/s} * 1\text{rev}/2\pi \text{ rad} = \boxed{0.20 \text{ rev/s}}$$

In this case, we didn't *need* to convert from revs to rads and back, since we never used a bridge equation. But better safe than sorry!

TASK 7.6

A rock is whirled at the end of a rope. If the length of the rope is halved, by what factor does the angular speed increase? The linear speed? The energy?

IT'S NOT A COLLISION OR AN EXPLOSION, BUT MAYBE WE CAN DO THIS USING ANGULAR MOMENTUM. THERE'S AN INITIAL AND A FINAL, SO IT'S WORTH A TRY!

R

α

ANGULAR-SPEED

$L = I\omega$; For a small object moving in a circle, $I = MR^2$, so
$$L = MR^2 \omega$$
Angular momentum is the same before and after the length of the rope is halved:
$$MR_i^2\omega_i = MR_f^2\omega_f$$
We're told that $R_f = \frac{1}{2} R_i$, so
$$MR_i^2\omega_i = M(\tfrac{1}{2} R_i)^2\omega_f$$
$$R_i^2\omega_i = 1/4 R_i^2 \omega_f$$
$$\boxed{\omega_f = 4 \omega_i}$$

LINEAR-SPEED

Linear speed = rω. if ω gets four times bigger and r is half as big, the linear speed is 4(1/2) = twice as big.

ENERGY

$KE = \frac{1}{2}mv^2$. If v gets twice as big and m stays the same, KE is 4 times bigger.

Or

$KE = \frac{1}{2}I\omega^2$. $I = mr^2$. If r is half as big, I is ¼ as big. If I is ¼ as big and ω is 4 times bigger, $\frac{1}{2}I\omega^2$ is $(1/4)(4)^2 = $ 4 times bigger.

IT ENDS WITH MORE ENERGY THAN IT STARTED WITH. THAT SEEMS WEIRD. WHERE DOES THE ENERGY COME FROM?

I KNOW WHERE! I CAN FEEL THIS ROCK PULLING ON ME THROUGH THE ROPE. THAT MEANS I'M PULLING ON IT. WE KNEW THAT ALREADY, SINCE IT'S MOVING IN A CIRCLE AND NEEDS A CENTRIPETAL FORCE. WHEN IT'S JUST GOING IN A CIRCLE, THAT FORCE IS PERPENDICULAR TO ITS MOTION, AND DOESN'T DO ANY WORK. BUT WHEN I PULL IT IN TO MAKE THE RADIUS SMALLER, THERE'S A PART OF ITS MOTION THAT IS PARALLEL TO THE TENSION, MEANING WORK IS DONE ON IT. SO THE ENERGY COMES FROM MY MUSCLES!

TASK 7.7

A uniform 26.0 m long board is balanced on a support at its center. When a 5000.0 kg triceratops sits on the left end of the board, she is balanced by a rock placed 4.0 m to the right of the support. What is the mass of the rock?

NOTHING IS MOVING AND THERE IS NO "INITIAL" AND "FINAL," SO I THINK WE NEED TO SOLVE THIS AS DYNAMICS WITH a = 0.

1 Choose the board as the object, since it connects everything.

2 Draw a free-body diagram. There is a force upward from the support and downward from the board's weight at the center. There are downward forces from Sadie and the rock.

I WILL DO A MINI-TASK TO UNDERSTAND THE FORCE I EXERT ON THE BOARD.

THE FORCES ON ME ARE $F_{BOARD\ ON\ SADIE}$ AND W_{SADIE}. SINCE I AM NOT ACCELERATING, THEY MUST BALANCE OUT: $F_{BOARD\ ON\ SADIE} = W_{SADIE}$. $F_{SADIE\ ON\ BOARD} = F_{BOARD\ ON\ SADIE}$, SO $F_{SADIE\ ON\ BOARD} = W_{SADIE}$! THE SAME SORT OF THING IS TRUE FOR THE ROCK.

$F_{BOARD\ ON\ SADIE}$

W_{SADIE}

Sadie just used a new principle from the books: Newton's Third Law. *The force one object exerts on a second is always equal in magnitude to the force the second object exerts on the first.*

Don't confuse that with forces cancelling because there's no acceleration. If the board were accelerating upward, for example, the force of the board on Sadie would have to be greater than her weight. And that upward force would, by Newton's Third Law, be equal to the force she exerts on the board, so the force she exerts on the board would be greater than her weight.

In this task nothing is accelerating, so the force she exerts on the board is just her weight.

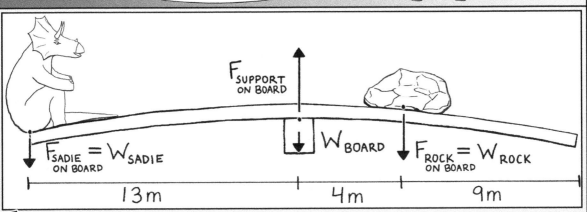

$$F_{\text{SUPPORT ON BOARD}}$$
$$F_{\text{SADIE ON BOARD}} = W_{\text{SADIE}}$$
$$W_{\text{BOARD}}$$
$$F_{\text{ROCK ON BOARD}} = W_{\text{ROCK}}$$

13m   4m   9m

3 There's no acceleration, so we can make either direction positive. Let's make up positive.

4 $\Sigma F = ma$
$-m_{\text{Sadie}}g + F_{\text{up}} - m_{\text{board}}g - m_{\text{rock}}g$
$= m_{\text{board}}a = m_{\text{board}}(0) = 0$

5 One equation, three unknowns.

Since forces alone aren't enough, Use the bridge equation to get torques from forces:
Sadie: $\tau = rF\sin\phi$
 $= (13.0)(5000.0)(9.8)\sin 90° = 637,000$
Support: $r = 0$, so $\tau = 0$
Board's weight: $r = 0$, so $\tau = 0$
Rock: $\tau = rF\sin\phi$
 $= (4.0)(m_{\text{rock}})(9.8)\sin 90° = 39.2\, m_{\text{rock}}$
Let's call torques that make the board go counterclockwise, like Sadie's, positive, and those that make it go clockwise negative.
$\Sigma\tau = I\alpha$.
$a = 0$, so $637,000 - 39.2\, m_{\text{rock}} = I_{\text{board}}(0) = 0$
$m_{\text{rock}} = 16,200\text{ kg}$.

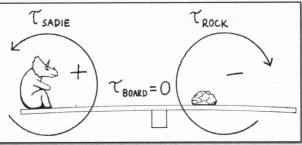

τ_{SADIE} τ_{ROCK}

$+$ $\tau_{\text{BOARD}} = 0$ $-$

WE COULD HAVE DONE THIS MORE EASILY IF WE'D STARTED WITH TORQUES INSTEAD OF FORCES.

YEAH, BUT I'M NOT SURE HOW TO TELL WHICH TO START WITH. IT DIDN'T HURT TO GO THROUGH THE FORCES TOO.

SOLUTION 7.8

1 Choose the board.

2 Draw a free-body diagram.

3 The board is not rotating, so we could choose either direction to be positive. Choose counterclockwise.

$\tau_{RIGHT} = 0$

F_{LEFT} F_{RIGHT} τ_{BOARD} τ_{MAIA}

τ_{LEFT}

W_{BOARD} W_{MAIA}

4A Use bridge equation to get torques.

$\tau_{\text{left support}} = -(2.00)F_{\text{left}}\sin 90° = -2.00 F_{\text{left}}$

$\tau_{\text{right support}} = 0$, since $r = 0$

$\tau_{\text{board}} = -(0.50)(0.60)(9.8)\sin 90° = -2.94$

$\tau_{\text{Maia}} = -(4.0)(2.0)(9.8)\sin 90° = -78.4$

4B

$\sum \tau = I\alpha = I(0) = 0$

$-2.00\, F_{\text{left}} - 2.94 - 78.4 = 0$

$F_{\text{left}} = -41\ N$

THE NEGATIVE TELLS US THE FORCE POINTS OPPOSITE HOW WE DREW IT IN THE FREE-BODY DIAGRAM. JUST LIKE OUR EXPERIMENT SHOWED!

WE STILL HAVE ONE SUPPORT FORCE TO FIND. I BET WE CAN DO IT EITHER BY CHOOSING ANOTHER PLACE TO MEASURE r FROM AND USING TORQUES, OR BY USING FORCES.

FORCES ARE PRETTY EASY ON THIS ONE. WE HAVE 41 N DOWN FROM THE LEFT SUPPORT, (0.60)(9.8) = 5.9 N DOWN FROM THE BOARD'S WEIGHT, AND (2.0)(9.8) = 19.6 N DOWN FROM MAIA'S WEIGHT. THE RIGHT SUPPORT HAS TO CANCEL ALL OF THOSE, SO IT NEEDS TO BE 41 + 5.9 + 19.6 = 66 N UP.

$F_{\text{RIGHT SUPPORT ON BOARD}}$ m_c

$F_{\text{LEFT SUPPORT ON BOARD}}$ W_{BOARD} W_{MAIA}

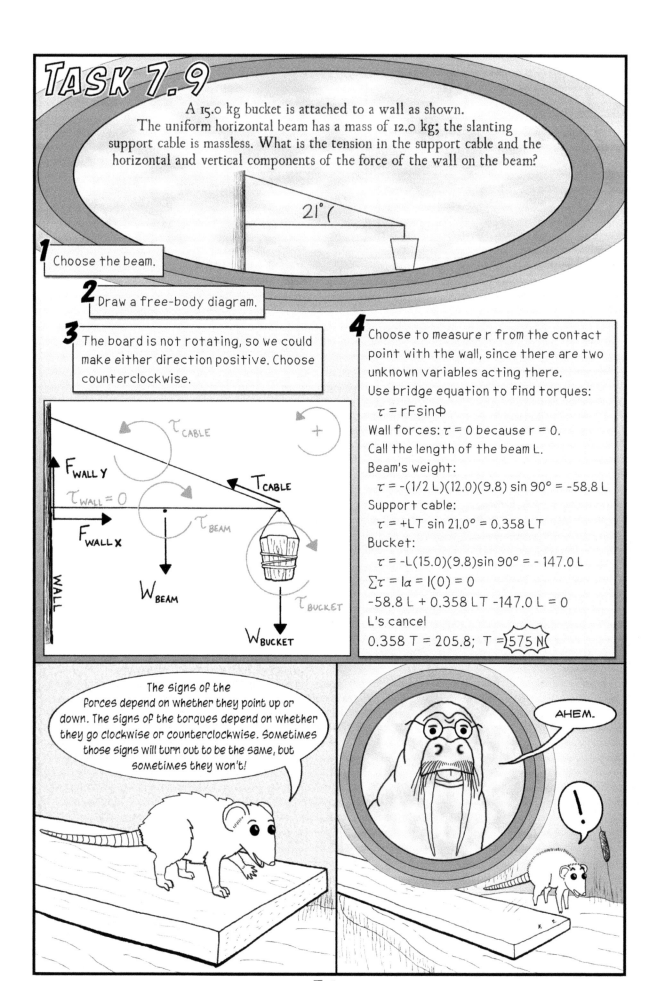

TASK 7.9

A 15.0 kg bucket is attached to a wall as shown. The uniform horizontal beam has a mass of 12.0 kg; the slanting support cable is massless. What is the tension in the support cable and the horizontal and vertical components of the force of the wall on the beam?

21°

1 Choose the beam.

2 Draw a free-body diagram.

3 The board is not rotating, so we could make either direction positive. Choose counterclockwise.

τ_{CABLE}

$+$

$F_{WALL\ Y}$

$\tau_{WALL} = 0$

$F_{WALL\ X}$

T_{CABLE}

τ_{BEAM}

WALL

W_{BEAM}

τ_{BUCKET}

W_{BUCKET}

4 Choose to measure r from the contact point with the wall, since there are two unknown variables acting there.
Use bridge equation to find torques:
$$\tau = rF\sin\phi$$
Wall forces: $\tau = 0$ because $r = 0$.
Call the length of the beam L.
Beam's weight:
$$\tau = -(1/2\ L)(12.0)(9.8)\sin 90° = -58.8\ L$$
Support cable:
$$\tau = +LT\sin 21.0° = 0.358\ LT$$
Bucket:
$$\tau = -L(15.0)(9.8)\sin 90° = -147.0\ L$$
$$\sum\tau = I\alpha = I(0) = 0$$
$$-58.8\ L + 0.358\ LT - 147.0\ L = 0$$
L's cancel
$$0.358\ T = 205.8;\quad T = \boxed{575\ N}$$

The signs of the forces depend on whether they point up or down. The signs of the torques depend on whether they go clockwise or counterclockwise. Sometimes those signs will turn out to be the same, but sometimes they won't!

AHEM.

!

76

KITSUNE'S ROTATIONAL PHYSICS CHART

Linear		Angular	Bridge
s	θ	angular displacement	$s = r\theta$
v	ω	angular speed	$v_t = r\omega$
a_t	α	angular acceleration	$a_t = r\alpha$
t	t	time	t
F	τ	torque	$\tau = rF\sin\theta$
m	I	moment of inertia	
p	L	angular momentum	L
KE	KE	rotational kinetic energy	

WORDS OF WISDOM

The first three bridge equations only work in radians

If something is rolling, the speed of its center of mass is the same as its tangential speed

Newton's Third Law: The force one object exerts on a second is always equal in magnitude to the force the second object exerts on the first

To calculate torques on a *non*-rotating object, choose an axis of rotation where there is an unknown force or distance

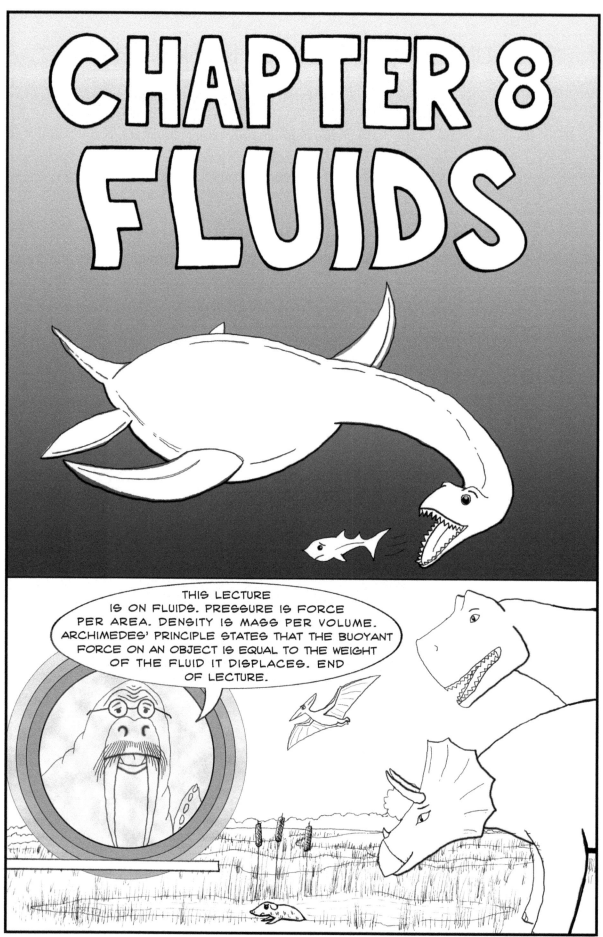

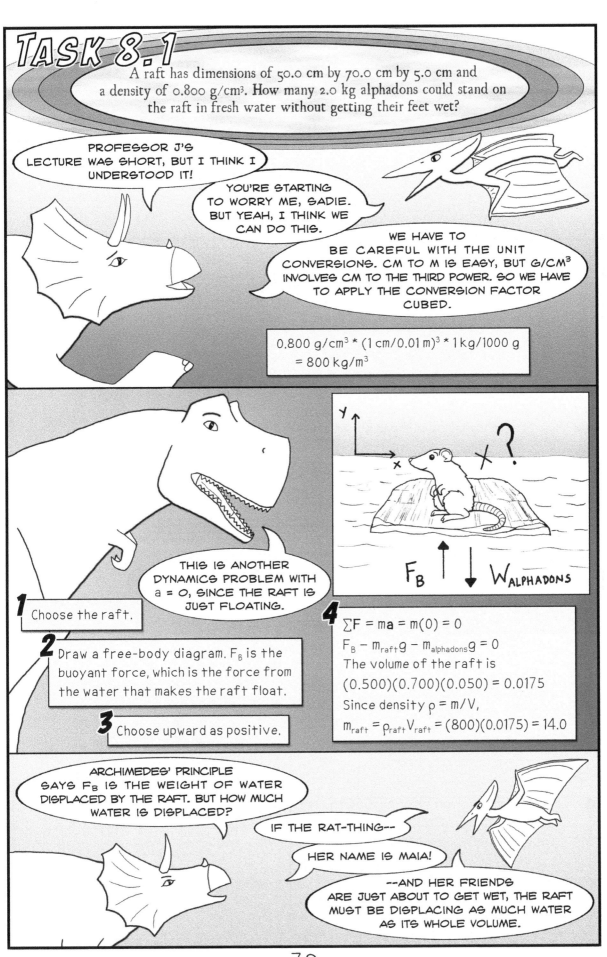

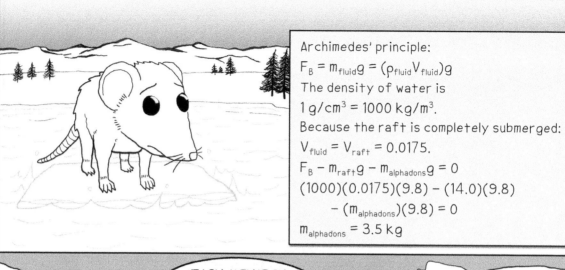

EACH ALPHADON IS 2.0 KG. 3.5/2.0 = 1.8 ALPHADONS.

BUT WE CAN'T HAVE A FRACTION OF AN ALPHADON.

WELL, NOT IF YOU WANT THEM ALIVE.

SO, SHOULD WE ROUND TO 2 ALPHADONS?

NO. IF 1.8 ALPHADONS IS THE LIMIT, 2 IS TOO MUCH. THE MAXIMUM NUMBER OF ALPHADONS THAT CAN FLOAT ON THE RAFT IS 1.

TASK 8.2

Ice has a density of 917 kg/m³. Salt water has a density of 1025 kg/m³. What fraction of an iceberg floating in salt water is under the surface?

1 Choose the iceberg.

2 Draw a free-body diagram.

3 $a = 0$. Choose upward as positive.

4 $\sum F = ma = m(0);\ \ F_B - m_{ice}g = 0$
$m_{saltwater}g - m_{ice}g = 0,\ \ g\text{'s cancel!}$
$m_{saltwater} = m_{ice}$
$\rho_{saltwater}V_{saltwater} = \rho_{ice}V_{ice}$
$V_{saltwater}/V_{ice} = \rho_{ice}/\rho_{saltwater} = 917/1025 = 0.89$
89% of the iceberg is under the surface.

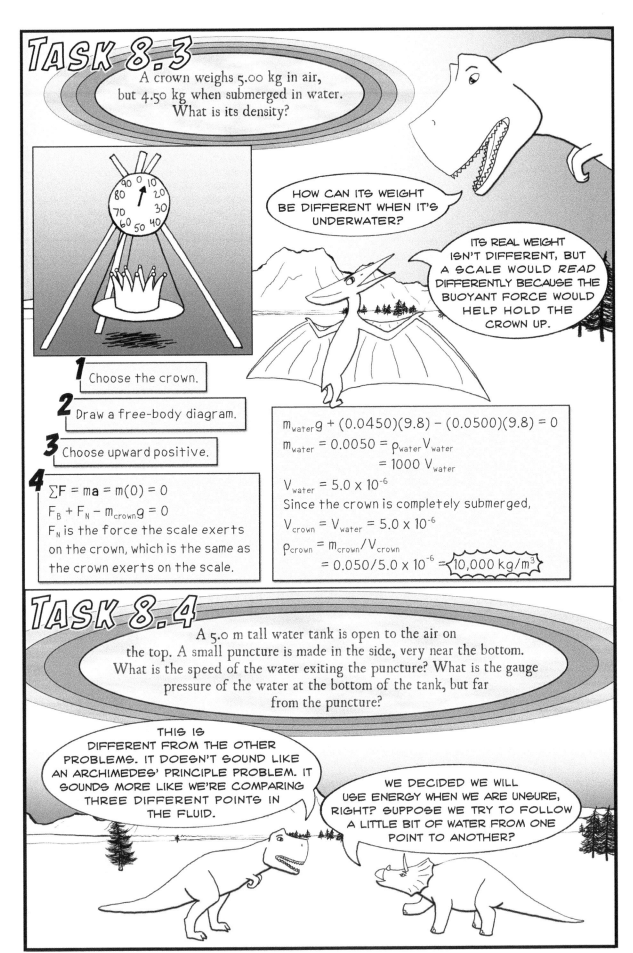

TASK 8.3

A crown weighs 5.00 kg in air, but 4.50 kg when submerged in water. What is its density?

1 | Choose the crown.

2 | Draw a free-body diagram.

3 | Choose upward positive.

4 |
$$\sum F = ma = m(0) = 0$$
$$F_B + F_N - m_{crown}g = 0$$
F_N is the force the scale exerts on the crown, which is the same as the crown exerts on the scale.

$$m_{water}g + (0.0450)(9.8) - (0.0500)(9.8) = 0$$
$$m_{water} = 0.0050 = \rho_{water} V_{water}$$
$$= 1000 V_{water}$$
$$V_{water} = 5.0 \times 10^{-6}$$
Since the crown is completely submerged,
$$V_{crown} = V_{water} = 5.0 \times 10^{-6}$$
$$\rho_{crown} = m_{crown}/V_{crown}$$
$$= 0.050/5.0 \times 10^{-6} = 10,000 \text{ kg/m}^3$$

HOW CAN ITS WEIGHT BE DIFFERENT WHEN IT'S UNDERWATER?

ITS REAL WEIGHT ISN'T DIFFERENT, BUT A SCALE WOULD *READ* DIFFERENTLY BECAUSE THE BUOYANT FORCE WOULD HELP HOLD THE CROWN UP.

TASK 8.4

A 5.0 m tall water tank is open to the air on the top. A small puncture is made in the side, very near the bottom. What is the speed of the water exiting the puncture? What is the gauge pressure of the water at the bottom of the tank, but far from the puncture?

THIS IS DIFFERENT FROM THE OTHER PROBLEMS. IT DOESN'T SOUND LIKE AN ARCHIMEDES' PRINCIPLE PROBLEM. IT SOUNDS MORE LIKE WE'RE COMPARING THREE DIFFERENT POINTS IN THE FLUID.

WE DECIDED WE WILL USE ENERGY WHEN WE ARE UNSURE, RIGHT? SUPPOSE WE TRY TO FOLLOW A LITTLE BIT OF WATER FROM ONE POINT TO ANOTHER?

THE QUESTION ASKS FOR **GAUGE PRESSURE**. WHAT'S THAT?

Compare the surface to a deep spot far from the puncture.

$P_{top} + 1/2\, \rho v_{top}^2 + \rho g h_{top}$
$\quad = P_{deep} + 1/2\, \rho v_{deep}^2 + \rho g h_{deep}$

Because the spot is far from the puncture, $v_{deep} \approx 0$.

$P_{top} + 1/2\,(1000)(0)^2 + (1000)(9.8)(5.0)$
$\quad = P_{deep} + 1/2\,(1000)(0)^2 + (1000)(9.8)(0)$

$P_{deep} = P_{top} + 49{,}000 \text{ Pa}$

IF I PUT A SCALE IN THE OPEN AIR, THE NEEDLE--THE GAUGE--READS ZERO EVEN THOUGH THERE'S MANY KG OF AIR PRESSING DOWN ON IT FROM THE ATMOSPHERE. I THINK IT'S LIKE THAT.

A PRESSURE GAUGE ONLY READS EXTRA PRESSURE ON IT BEYOND THE ATMOSPHERIC PRESSURE, SO THAT EXTRA PRESSURE IS CALLED THE GAUGE PRESSURE. IN THIS TASK, P_{TOP} IS ATMOSPHERIC, SO THE EXTRA PRESSURE AT THE BOTTOM IS 49,000 PA, AND THAT'S OUR ANSWER.

TASK 8.5

A pipe carries water from street level to the second-floor apartment, 4.00 m above. The faucet on the top floor has three-quarters the diameter of the street-level pipe. If we want the water to come out of the faucet at 8.00 m/s, what is the required gauge pressure in the street level pipe, assuming no other taps are open?

$P_{top} + 1/2\, \rho v_{top}^2 + \rho g h_{top}$
$\quad = P_{street} + 1/2\, \rho v_{street}^2 + \rho g h_{street}$

$P_{top} + 1/2\,(1000)(8.00)^2 + (1000)(9.8)(4.00)$
$\quad = P_{street} + 1/2(1000)v_{street}^2 + (100)(9.8)(0)$

$P_{top} + 71{,}200 = P_{street} + 500\, v_{street}^2$

WHAT IS V_{street}? I HAVE A FEELING THAT IT RELATES TO THE DIAMETER OF THE FAUCET SOMEHOW.

LET'S FIND OUT. SIMPLICIO, WOULD YOU LIKE TO STAMPEDE A HERD OF VELOCIRAPTORS?

I LOVE STAMPEDING! I'M ON IT!

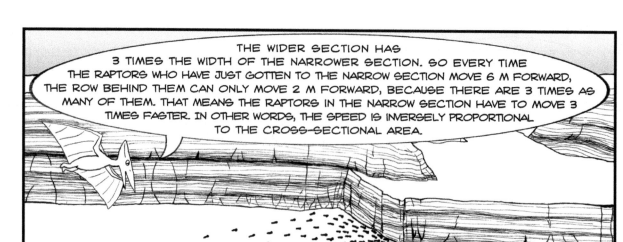

THE WIDER SECTION HAS 3 TIMES THE WIDTH OF THE NARROWER SECTION. SO EVERY TIME THE RAPTORS WHO HAVE JUST GOTTEN TO THE NARROW SECTION MOVE 6 M FORWARD, THE ROW BEHIND THEM CAN ONLY MOVE 2 M FORWARD, BECAUSE THERE ARE 3 TIMES AS MANY OF THEM. THAT MEANS THE RAPTORS IN THE NARROW SECTION HAVE TO MOVE 3 TIMES FASTER. IN OTHER WORDS, THE SPEED IS INVERSELY PROPORTIONAL TO THE CROSS-SECTIONAL AREA.

The textbooks call this the *equation of continuity*. It applies to incompressible fluids, like water, but not compressible fluids, like air.

SOLUTION 8.5 CONTINUED

$P_{top} + 71,200 = P_{street} + 500(4.5)^2$

$P_{street} = P_{top} + 61,100$ Pa.

The gauge pressure at street level is 61,100 Pa.

TASK 8.6

The stem of a brake pedal in a car with manual brakes has a radius of 0.500 cm. The disc pads themselves have an area of 60.0 cm² per wheel. If a driver slowly applies 100.0 N to the brake pedal, how much force is applied to each of the four wheels by the hydraulic fluid? The wheels are at the same height as the brake pedal.

THIS IS ANOTHER PROBLEM REFERRING TO TECHNOLOGY OF THE DISTANT FUTURE! WE CAN SOLVE THESE WITHOUT UNDERSTANDING ALL THE DETAILS OF THAT TECHNOLOGY.

$P_{pedal} + 1/2\,\rho v_{pedal}^2 + \rho g h_{pedal}$
$= P_{pad} + 1/2\,\rho v_{pad}^2 + \rho g h_{pad}$

Since they're at the same height, the ρgh terms cancel, and since the motion is slow, v ≈ 0.

So $P_{pad} = P_{pedal} = F_{pedal}/A_{pedal}$
$= 100.0/[\pi(0.00500)^2] = 1.27 \times 10^6$

Each brake pad has an area of 60.0 cm² × (0.01 m/cm)² = 0.00600 m².

Since there are four of them, the total area of the pads is 4(0.00600) = 0.0240.

$1.27 \times 10^6 = F_{pad}/A_{pad} = F_{pad}/0.0240$

$F_{pad} = 30,500$ N.

EQUATIONS IN THIS CHAPTER

$$P = F/A$$

$$\rho = m/V$$

$$F_B = \text{weight of fluid displaced}$$

$$P + 1/2\,\rho v^2 + \rho g h = \text{constant}$$

$$vA = \text{constant (incompressible fluid)}$$

CHAPTER 9
SIMPLE HARMONIC MOTION

IT IS MY TURN TO PROVIDE A LECTURE. MY POSITION IS STILL PROVISIONAL, AND MY TEACHING IS BEING OBSERVED.

THAT MEANS SHE CAN'T SEEM TOO HELPFUL.

MOST SMALL OSCILLATIONS, WHETHER OF A MASS ON THE END OF A SPRING, A PENDULUM SWINGING BACK AND FORTH, OR A BOAT BOBBING ON THE WATER, FOLLOW THE SAME MATHEMATICAL PATTERN.

THE KINEMATICS IS GIVEN BY THE EQUATION OF MOTION $\Delta x = A\cos(\omega t)$, WHERE A IS THE **AMPLITUDE** OF THE MOTION AND ω THE **ANGULAR FREQUENCY**. FOR A MASS m ON THE END OF A SPRING, $\omega = \sqrt{(k/m)}$. FOR A SIMPLE PENDULUM, $\omega = \sqrt{(g/L)}$, WHERE L IS THE LENGTH OF THE PENDULUM. THE MAGNITUDE OF THE FORCE OF A SPRING IS GIVEN BY $F = kx$. k IS THE **SPRING CONSTANT** AND MEASURES THE STIFFNESS OF THE SPRING.

FROM THESE FACTS, AND FROM WHAT YOU ALREADY KNOW, YOU CAN FIGURE OUT THE REST.

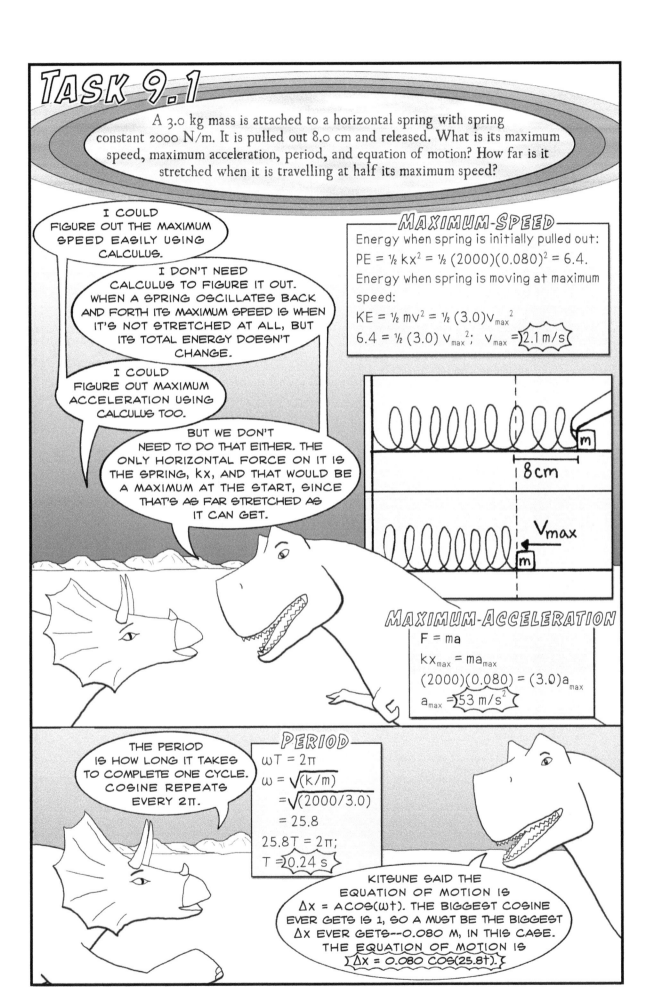

FOR HOW STRETCHED IT IS WHEN TRAVELLING AT HALF ITS MAXIMUM SPEED, WE CAN USE ENERGY AGAIN.

STRETCH AT HALF MAXIMUM SPEED

Energy when spring is initially pulled out:

$PE = \frac{1}{2} kx^2 = \frac{1}{2}(2000)(0.080)^2 = 6.4$.

Energy when travelling at half-maximum speed is a mix of PE and KE:

$\frac{1}{2} kx^2 + \frac{1}{2} m(\frac{1}{2} v_{max})^2 = 6.4$

$\frac{1}{2}(2000)x^2 + \frac{1}{2}(3.0)(2.1/2)^2 = 6.4$

$1000x^2 + 1.7 = 6.4; \quad x = 0.069\ m$

Writing the equation of motion using cosine assumes the motion starts all the way stretched out, like it does here. If it starts in the middle, say by giving the mass a tap, the equation of motion would use sine.

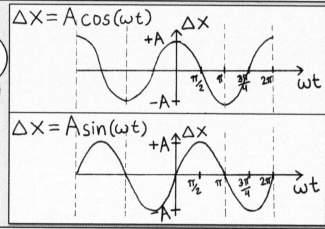

$\Delta X = A\cos(\omega t)$

$\Delta X = A\sin(\omega t)$

TASK 9.2

You notice that your boat is oscillating up and down on the water according to simple harmonic motion with a period of 12.0 seconds and an amplitude of 0.25 m. Find the maximum speed and acceleration of the boat's oscillation.

KITSUNE SAID ALL THESE OSCILLATIONS FOLLOW THE SAME MATHEMATICAL PATTERN. SO EVEN THOUGH IT IS NOT A MASS ON THE END OF A SPRING WE CAN SOLVE IT AS IF IT WERE.

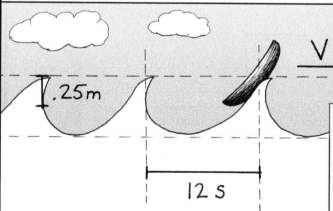

.25m

12 s

Energy at top of motion: $\frac{1}{2} kx_{max}^2$

Energy at maximum speed: $\frac{1}{2} mv_{max}^2$

$\frac{1}{2} kx_{max}^2 = \frac{1}{2} mv_{max}^2$

$v_{max}^2 = k/m\ x_{max}^2$

$v_{max} = \sqrt{(k/m)}\ x_{max}$

I RECOGNIZE $\sqrt{k/m}$. THAT'S THE ANGULAR FREQUENCY ω! AND AMPLITUDE MEANS THE SAME THING AS x_{max}.

$v_{max} = \sqrt{(k/m)}\ x_{max}$; $v_{max} = \omega A$

Period = 12, so $\omega T = 2\pi$

$\omega(12) = 2\pi$; $\omega = 0.524$

$v_{max} = (0.524)(0.25) = 0.13$ m/s

$F = kx = ma$

$kx_{max} = ma_{max}$

$a_{max} = k/m\ x_{max} = \omega^2 A$

$\quad = (0.524)^2(0.25) = 0.069$ m/s^2

TASK 9.3

When a 2.0 kg mass is suspended from a spring, the spring stretches 13.0 cm. If it is stretched 3.0 cm more and then released, what is the maximum speed of the resulting oscillation?

WHEN THE MASS IS SUSPENDED FROM THE SPRING, IT'S NOT OSCILLATING YET, BUT IT'S IN BALANCE. THE WEIGHT DOWN BALANCES THE SPRING FORCE UP.

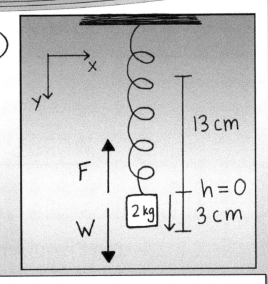

X

Y

F

W

13 cm

h = 0

3 cm

2 kg

$mg = kx$

$(2.0)(9.8) = k(0.130)$; $k = 150.8$

Choose the height at which it is in balance as h = 0.

Initially, it's stretched 16.0 cm total and it's at a height of -3.0 cm relative to h = 0.

So $E_i = \frac{1}{2} kx^2 + mgh = \frac{1}{2}(150.8)(0.160)^2 + (2.0)(9.8)(-0.030) = 1.34$

Its maximum speed will be when it passes through that balance point (h = 0), so

$E_f = \frac{1}{2} kx^2 + mgh + \frac{1}{2} mv_{max}^2 = \frac{1}{2}(150.8)(0.130)^2 + (2.0)(9.8)(0) + \frac{1}{2}(2.0) v_{max}^2$

$\quad = 1.27 + v_{max}^2 = 1.34$

$v_{max} = 0.26$ m/s

THERE'S AN EASIER WAY TO DO IT IF WE TRUST KITSUNE. SHE SAID THE MATHEMATICAL PATTERN IS THE SAME FOR ALL SMALL OSCILLATIONS. SO THIS SHOULD HAVE THE SAME PATTERN AS A HORIZONTAL SPRING. IF WE FIND THE SPRING CONSTANT THE WAY SIMPLICIO DID, THEN WE CAN PRETEND THE SPRING IS HORIZONTAL AND IS STRETCHED 3.0 CM.

EQUATIONS IN THIS CHAPTER

$$\Delta x = A\cos(\omega t) \quad \text{(small oscillations)}$$
$$\omega = \sqrt{(k/m)} \quad \text{(mass on spring)}$$
$$\omega = \sqrt{(g/L)} \quad \text{(pendulum)}$$
$$F = kx \quad \text{(Force of a spring)}$$

WORDS OF WISDOM

All small oscillations follow the same
mathematical pattern

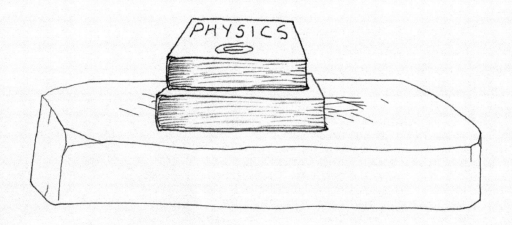

93

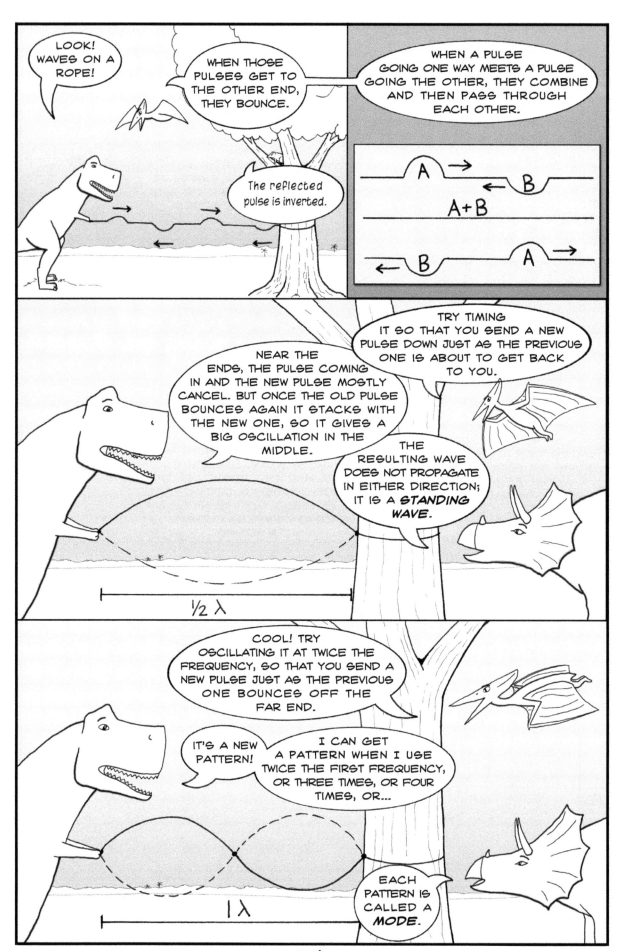

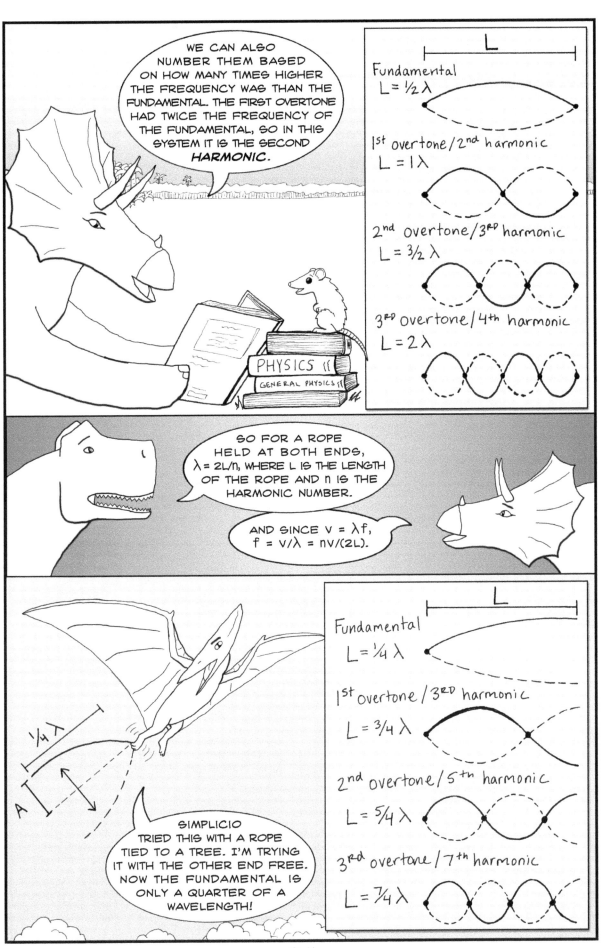

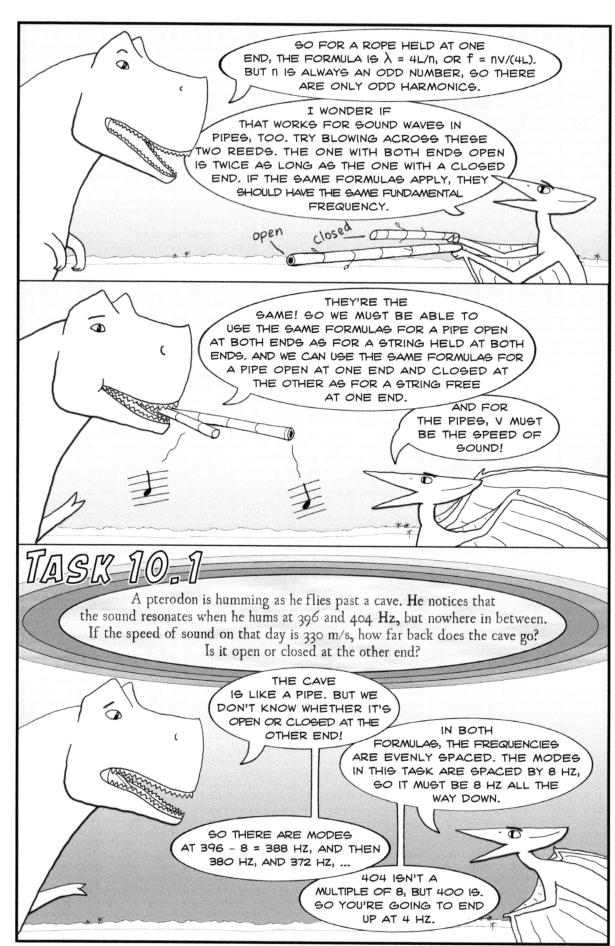

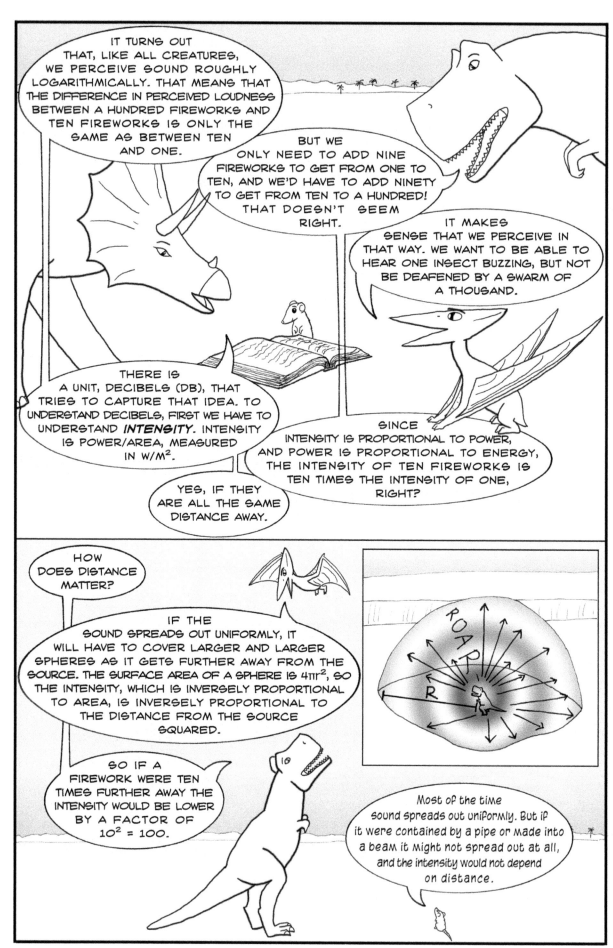

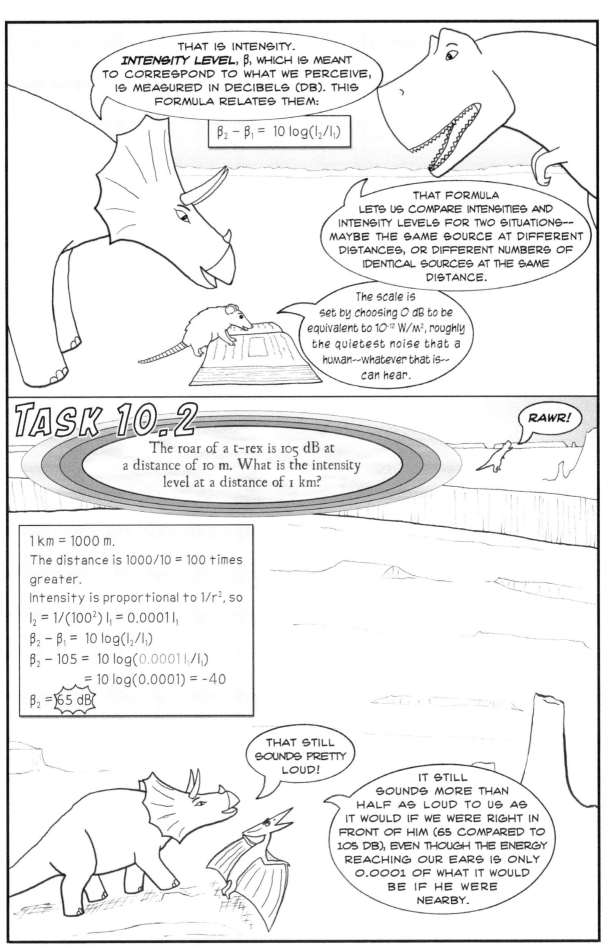

404
} 8
396
} 8
388
} 8
380
} 8
372
...
12 1st overtone
= 3rd harmonic
} 8
4 Fundamental!
SO NOW WE KNOW THE FORMULA IS f = nv/(4L)! AND THE FUNDAMENTAL FREQUENCY IS 4 HZ!
SINCE WE ONLY HAVE ODD HARMONICS, THE CAVE IS CLOSED AT THE OTHER END!
f = nv/4L
4 = 1(330)/(4*L)
16L = 330; L = 21 m
I'M GOING TO CELEBRATE BY SETTING OFF SOME OF TERRANCE'S FIREWORKS!
THEY ARE REALLY LOUD!
YEAH! I'M GOING TO TRY TEN AT ONCE!
OH NO! THAT WILL BE TEN TIMES AS LOUD!
HM. THAT SOUNDED A LITTLE LOUDER, BUT NOT TEN TIMES AS LOUD. IS THERE SOMETHING IN ONE OF THOSE BOOKS ABOUT THIS, LITTLE ONE?

THAT IS INTENSITY. **INTENSITY LEVEL**, β, WHICH IS MEANT TO CORRESPOND TO WHAT WE PERCEIVE, IS MEASURED IN DECIBELS (DB). THIS FORMULA RELATES THEM:

$$\beta_2 - \beta_1 = 10 \log(I_2/I_1)$$

THAT FORMULA LETS US COMPARE INTENSITIES AND INTENSITY LEVELS FOR TWO SITUATIONS-- MAYBE THE SAME SOURCE AT DIFFERENT DISTANCES, OR DIFFERENT NUMBERS OF IDENTICAL SOURCES AT THE SAME DISTANCE.

The scale is set by choosing 0 dB to be equivalent to 10^{-12} W/M², roughly the quietest noise that a human--whatever that is-- can hear.

TASK 10.2

The roar of a t-rex is 105 dB at a distance of 10 m. What is the intensity level at a distance of 1 km?

RAWR!

1 km = 1000 m.
The distance is 1000/10 = 100 times greater.
Intensity is proportional to $1/r^2$, so
$I_2 = 1/(100^2)\, I_1 = 0.0001\, I_1$
$\beta_2 - \beta_1 = 10 \log(I_2/I_1)$
$\beta_2 - 105 = 10 \log(0.0001\, I_1/I_1)$
$\qquad\qquad = 10 \log(0.0001) = -40$
$\beta_2 = 65\ dB$

THAT STILL SOUNDS PRETTY LOUD!

IT STILL SOUNDS MORE THAN HALF AS LOUD TO US AS IT WOULD IF WE WERE RIGHT IN FRONT OF HIM (65 COMPARED TO 105 DB), EVEN THOUGH THE ENERGY REACHING OUR EARS IS ONLY 0.0001 OF WHAT IT WOULD BE IF HE WERE NEARBY.

EQUATIONS IN THIS CHAPTER

$$v = \lambda f$$

both ends the same: $f = nv/(2L)$ $n = 1, 2, 3, \ldots$

ends are different: $f = nv/(4L)$ $n = 1, 3, 5, \ldots$

$$\beta_2 - \beta_1 = 10 \log(I_2/I_1)$$

WORDS OF WISDOM

Overtones are numbered in sequence; harmonics are multiples of the fundamental frequency

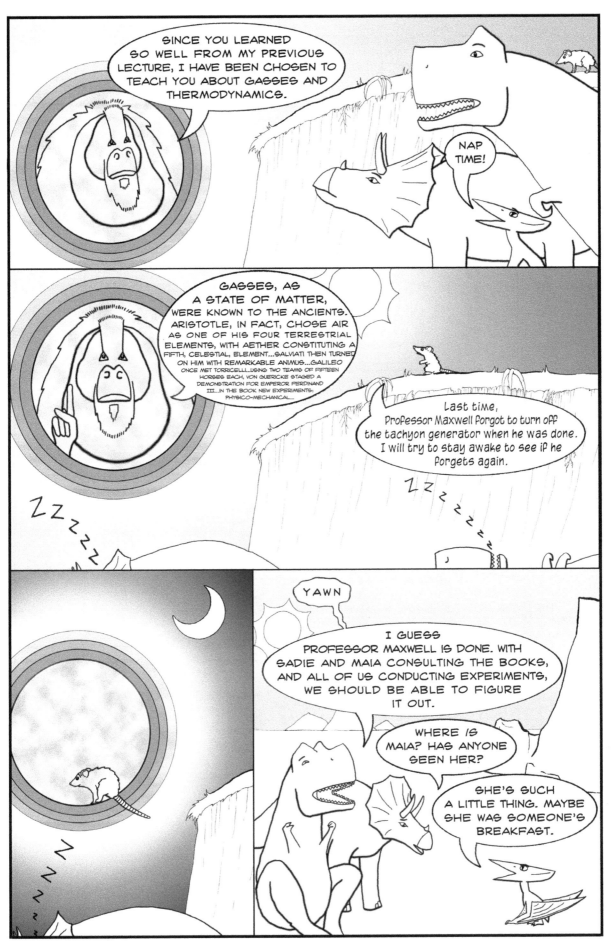

IN ORDER TO COMPUTE THIS, WE BREAK IT INTO STEPS. FOR EACH STEP WITH A TEMPERATURE CHANGE, WE USE $q = mc\Delta T$. q IS HEAT, m IS MASS, c IS SPECIFIC HEAT FROM A TABLE, AND ΔT IS CHANGE IN TEMPERATURE. FOR EACH STEP WITH A PHASE CHANGE, LIKE FROM SOLID TO LIQUID OR LIQUID TO GAS, WE USE $q = +/- mL$, WHERE L IS THE LATENT HEAT, ALSO FROM A TABLE. IT'S + IF WE'RE ADDING HEAT AND − IF WE'RE TAKING IT AWAY.

THE SECOND PART IS EASY, BECAUSE POWER IS ENERGY TRANSFERRED PER TIME. HEAT IS A FORM OF ENERGY TRANSFER, SO WE CAN JUST WRITE $P = q/t$, OR $t = q/P$.

1 Raise temperature of ice from −20°C to 0°C, the melting point of ice.
$q = mc\Delta T$
$m = (30.0\ g)(1\ kg/1000\ g) = 0.0300\ kg$
$c_{ice} = 2050\ J\,kg^{-1}\,°C^{-1}$
$q = (0.0300)(2050)(0 - -20) = 1230\ J$
$t = q/P = 1220/300 = $ 4 s

2 Melt ice.
$L_{fusion,\ ice\ to\ water} = 3.34 \times 10^5\ J/kg$
$q = +mL = (0.0300)(3.34 \times 10^5) = 10,000\ J$
$t = q/P = 10000/300 = $ 33 s

3 Raise temperature of water from 0°C to 100°C, the boiling point of water.
$q = mc\Delta T$
$c_{water} = 4180\ J\,kg^{-1}\,°C^{-1}$
$q = (0.0300)(4180)(100 - 0) = 12,500\ J$
$t = q/P = 12500/300 = $ 42 s

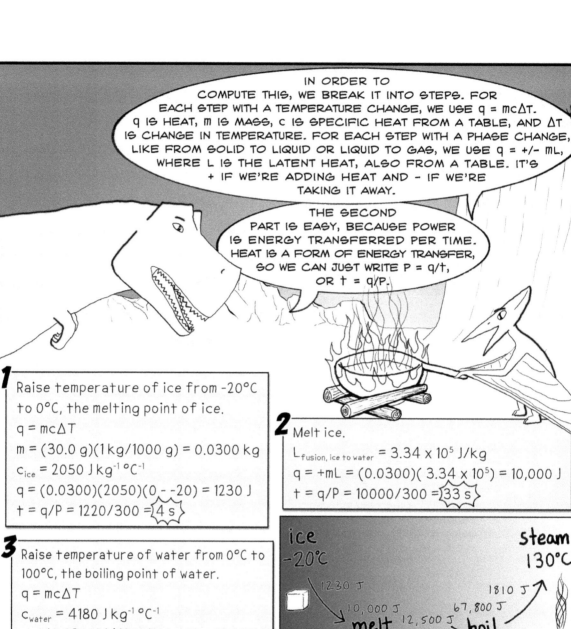

ice −20°C
1230 J
melt 0°C
10,000 J
12,500 J
boil 100°C
67,800 J
1810 J
steam 130°C

4 Boil water.
$L_{vaporization,\ water\ to\ steam} = 2.26 \times 10^6\ J/kg$
$q = +mL$
$\quad = (0.0300)(2.26 \times 10^6) = 67,800\ J$
$t = q/P = 67800/300 = $ 226 s

5 Raise temperature of steam from 100°C to 130°C.
$q = mc\Delta T$
$c_{steam} = 2010\ J\,kg^{-1}\,°C^{-1}$
$q = (0.0300)(2010)(130 - 100) = 1810\ J$
$t = q/P = 1810/300 = $ 6 s

JUST ADD UP TO GET THE TOTAL HEAT NEEDED:
$1230 + 10,000 + 12,500 + 67,800 + 1810 = $ 93,300 J.

MORE ENERGY WENT IN TO CHANGING THE WATER TO STEAM THAN ALL THE OTHER STEPS PUT TOGETHER.

300 g of iron at 400°C is dropped into 150 g of water in a 120 g copper calorimeter, both initially at 17°C. Assuming no heat is lost to the environment, what is the final temperature of the system?

I SEE HOW TO DO THIS ONE!

IF NO HEAT IS LOST TO THE ENVIRONMENT THEN mcΔT FOR EACH OF THE THREE SUBSTANCES HAS TO ADD TO ZERO.

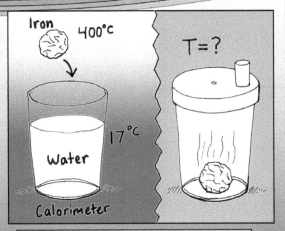

Iron 400°C

Water

17°C

Calorimeter

T=?

$c_{iron} = 450 \ J\,kg^{-1}\,°C^{-1}$
$q_{iron} = mc\Delta T = (0.300)(450)(T_f - 400)$

$c_{water} = 4180 \ J\,kg^{-1}\,°C^{-1}$
$q_{water} = mc\Delta T = (0.150)(4180)(T_f - 17)$

$c_{copper} = 390 \ J\,kg^{-1}\,°C^{-1}$
$q_{copper} = mc\Delta T = (0.120)(390)(T_f - 17)$

$\sum q = 0$
$(0.300)(450)(T_f - 400)$
$\quad + (0.150)(4180)(T_f - 17)$
$\quad + (0.120)(390)(T_f - 17) = 0$
$135 \ T_f - 54,000 + 627 \ T_f - 10,700$
$\quad + 47 T_f - 800 = 0$
$809 \ T_f = 65,500; \quad T_f = \boxed{81°C}$

50 g of ice at −5°C is dropped into 500 g of water at 33°C. Assuming the system comes to equilibrium quickly, what is the final temperature and phase of the system?

SINCE IT COMES TO EQUILIBRIUM QUICKLY, NOT MUCH HEAT IS EXCHANGED WITH THE ENVIRONMENT, SO THE SUM OF THE HEATS IS ZERO AGAIN.

BUT WE DO NOT KNOW WHETHER IT ENDS UP AS ALL ICE, PARTIALLY MELTED ICE IN WATER, OR ALL WATER.

NO WAY IT'S GOING TO END UP AS ALL ICE. THERE'S LESS ICE THAN WATER, AND THE ICE IS CLOSER TO 0°C THAN THE WATER IS. THE ICE WILL END UP EITHER PARTIALLY MELTING OR WE'LL END UP WITH ALL WATER.

LET'S GUESS! I'LL GUESS IT PARTIALLY MELTS. WE'LL SOLVE FOR THE AMOUNT THAT MELTS, AND IF WE'RE WRONG WE'LL END UP WITH AN IMPOSSIBLE NUMBER--EITHER MORE ICE THAN THERE IS OR A NEGATIVE AMOUNT.

If the ice partially melts, the final temperature is 0°C.

Heat the ice from -5 to 0°C:
$q = mc\Delta T$
$\quad = (0.050)(2050)(0 - -5) = 512$

Cool the water from 33 to 0°C:
$q = mc\Delta T$
$\quad = (0.500)(4180)(0 - 33) = -69{,}000$

Partially melt a mass m of the ice:
$q = +mL = m(3.34 \times 10^5)$

$\sum q = 0$
$512 - 69{,}000 + m(3.34 \times 10^5) = 0$
$(3.34 \times 10^5)m = 68{,}400$
$m = 0.205\,kg = 205\,g$

$m_{ice} = ?$ partially melts

ice -5°C   water 33°C

$T_f = 0°C$

THAT'S MORE ICE THAN THERE WAS! YOU GUESSED WRONG. I'LL TRY IT WITH ALL THE ICE MELTING.

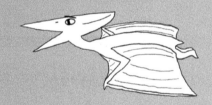

Heat the ice from -5 to 0°C:
$q = mc\Delta T$
$\quad = (0.050)(2050)(0 - -5) = 512$

Melt all the ice:
$q = +mL$
$\quad = (0.050)(3.34 \times 10^5) = 16{,}700$

Heat the melted ice from 0°C to T_f:
$q = mc\Delta T = (0.050)(4180)(T_f - 0)$
$\quad = 209T_f$

Cool the water from 33°C to T_f:
$q = mc\Delta T = (0.500)(4180)(T_f - 33)$
$\quad = 2090T_f - 69{,}000$

$\sum q = 0$
$512 + 16{,}700 + 209T_f + 2090T_f - 69{,}000 = 0$
$2300T_f = 51{,}800;\quad T_f = 23°C \text{ (all water)}$

$m_{ice} = 0$ all melts

ice -5°C   water 33°C

$T_f = ?$

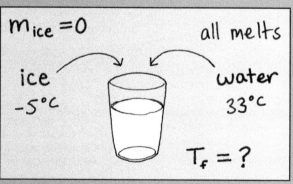

IF I'D GUESSED WRONG, I'D HAVE GOTTEN A TEMPERATURE BELOW 0°C.

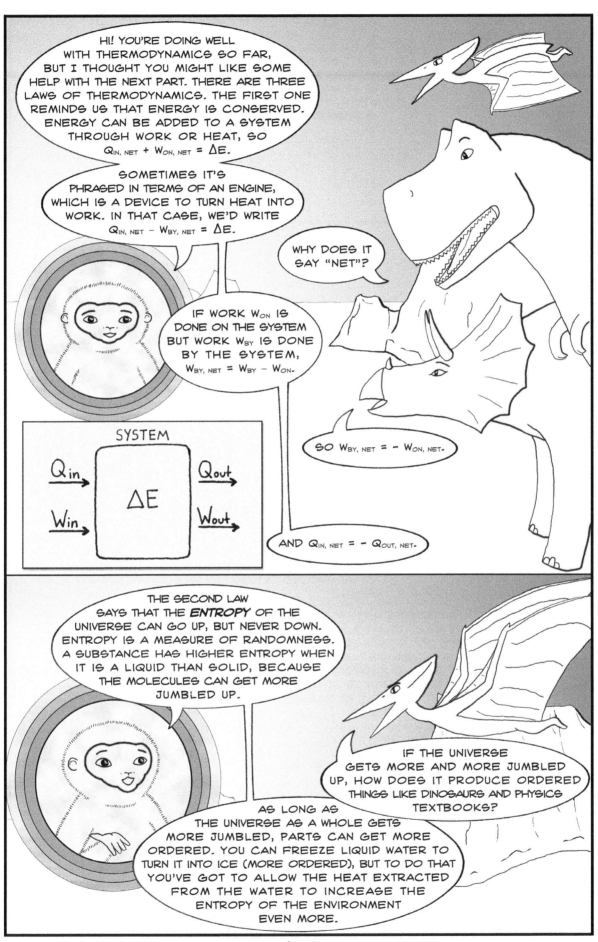

109

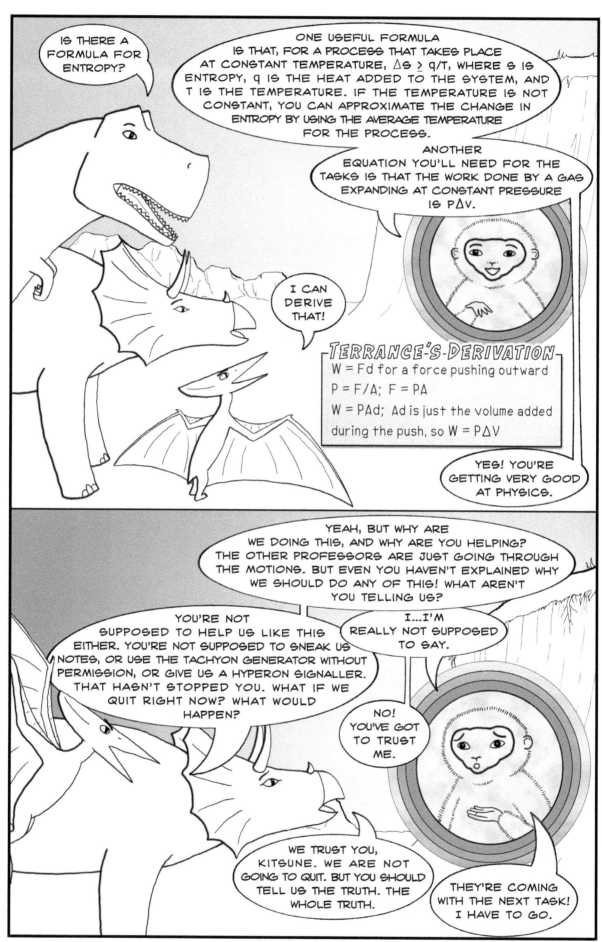

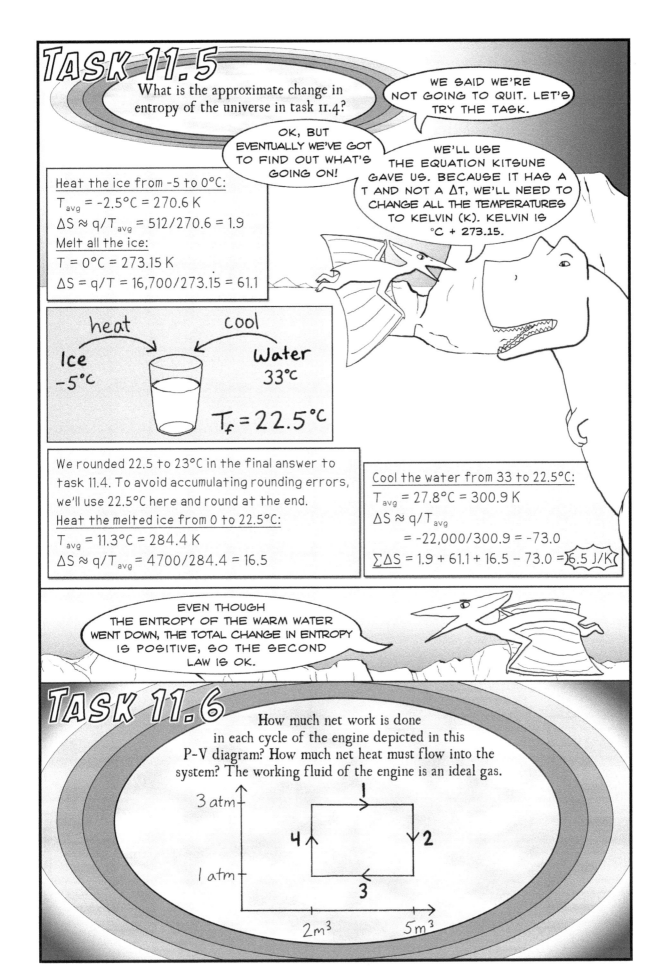

TASK 11.5

What is the approximate change in entropy of the universe in task 11.4?

WE SAID WE'RE NOT GOING TO QUIT. LET'S TRY THE TASK.

OK, BUT EVENTUALLY WE'VE GOT TO FIND OUT WHAT'S GOING ON!

WE'LL USE THE EQUATION KITSUNE GAVE US. BECAUSE IT HAS A T AND NOT A ΔT, WE'LL NEED TO CHANGE ALL THE TEMPERATURES TO KELVIN (K). KELVIN IS °C + 273.15.

Heat the ice from -5 to 0°C:
$T_{avg} = -2.5°C = 270.6\ K$
$\Delta S \approx q/T_{avg} = 512/270.6 = 1.9$
Melt all the ice:
$T = 0°C = 273.15\ K$
$\Delta S = q/T = 16{,}700/273.15 = 61.1$

heat → Ice -5°C

cool → Water 33°C

$T_f = 22.5°C$

We rounded 22.5 to 23°C in the final answer to task 11.4. To avoid accumulating rounding errors, we'll use 22.5°C here and round at the end.
Heat the melted ice from 0 to 22.5°C:
$T_{avg} = 11.3°C = 284.4\ K$
$\Delta S \approx q/T_{avg} = 4700/284.4 = 16.5$

Cool the water from 33 to 22.5°C:
$T_{avg} = 27.8°C = 300.9\ K$
$\Delta S \approx q/T_{avg}$
$= -22{,}000/300.9 = -73.0$
$\sum \Delta S = 1.9 + 61.1 + 16.5 - 73.0 = 6.5\ J/K$

EVEN THOUGH THE ENTROPY OF THE WARM WATER WENT DOWN, THE TOTAL CHANGE IN ENTROPY IS POSITIVE, SO THE SECOND LAW IS OK.

TASK 11.6

How much net work is done in each cycle of the engine depicted in this P-V diagram? How much net heat must flow into the system? The working fluid of the engine is an ideal gas.

3 atm

1 atm

4

1

2

3

2m³   5m³

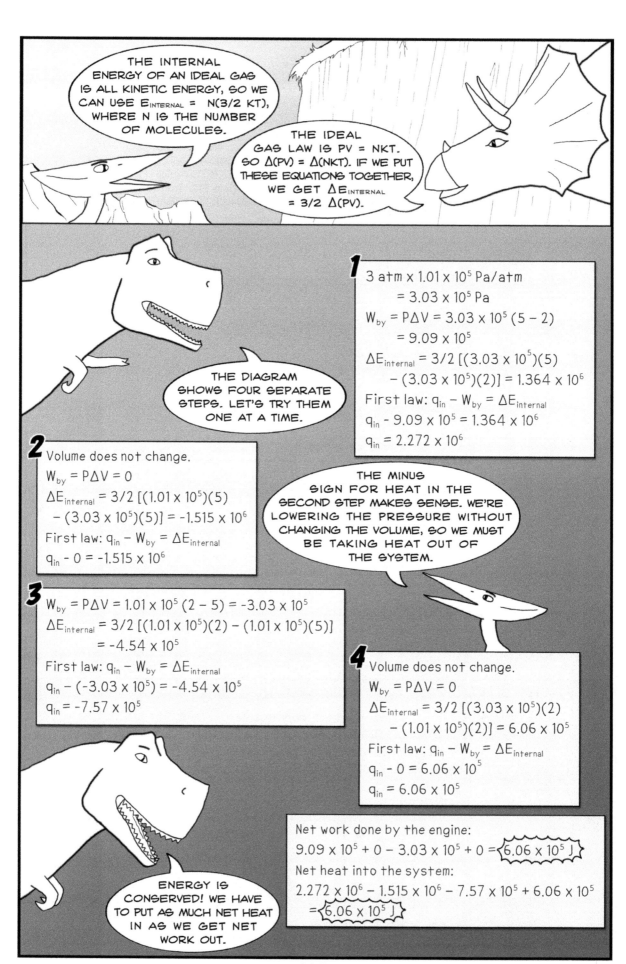

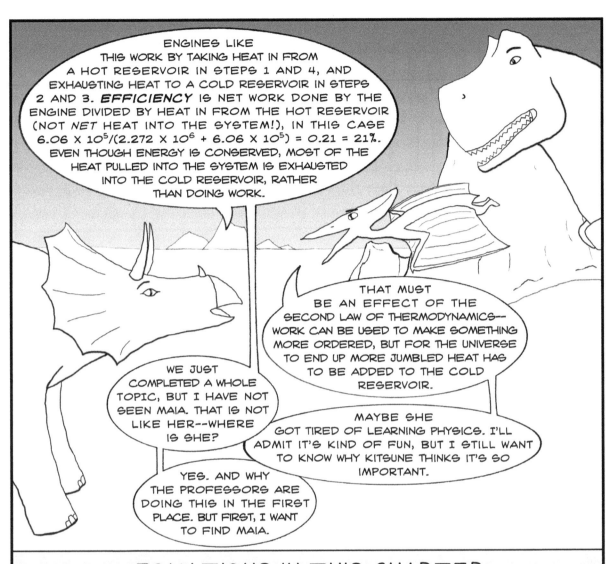

EQUATIONS IN THIS CHAPTER

$$Q_{in,\,net} + W_{on,\,net} = \Delta E$$

$$q = mc\Delta T; \quad q_{phase\ change} = \pm mL$$

$$W_{by\ a\ gas} = P\Delta V$$

$$\Delta S \geq q/T$$

WORDS OF WISDOM

The entropy of something can decrease as long as the total entropy of the universe increases

115

Will Maia make it home?

Will the dinosaurs avoid extinction?

Will Kitsune get tenure?

And what is an electric field, anyway?

Stay tuned for more exciting physics adventures!

CHAPTER 12
ELECTRIC
FIELDS

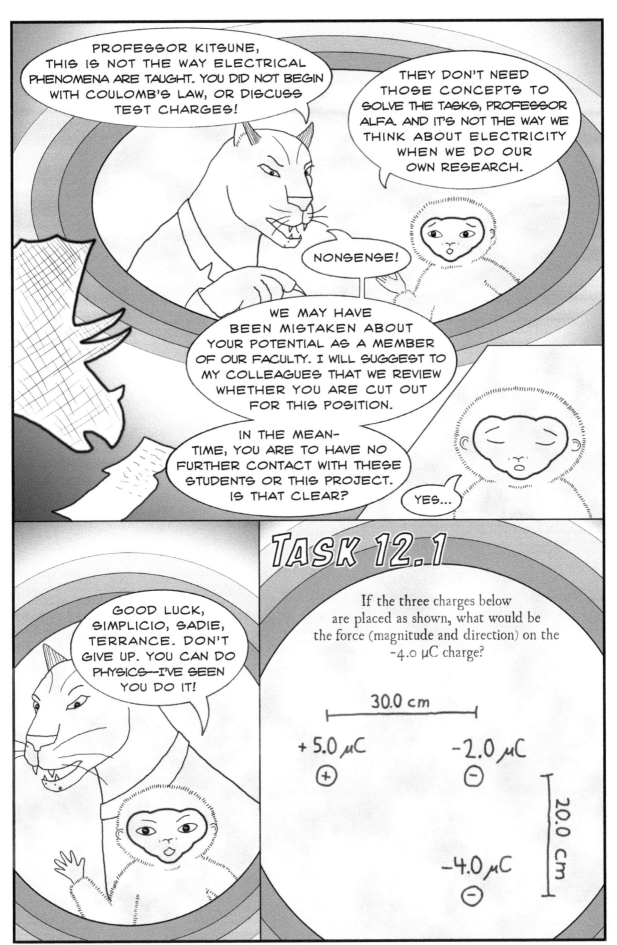

119

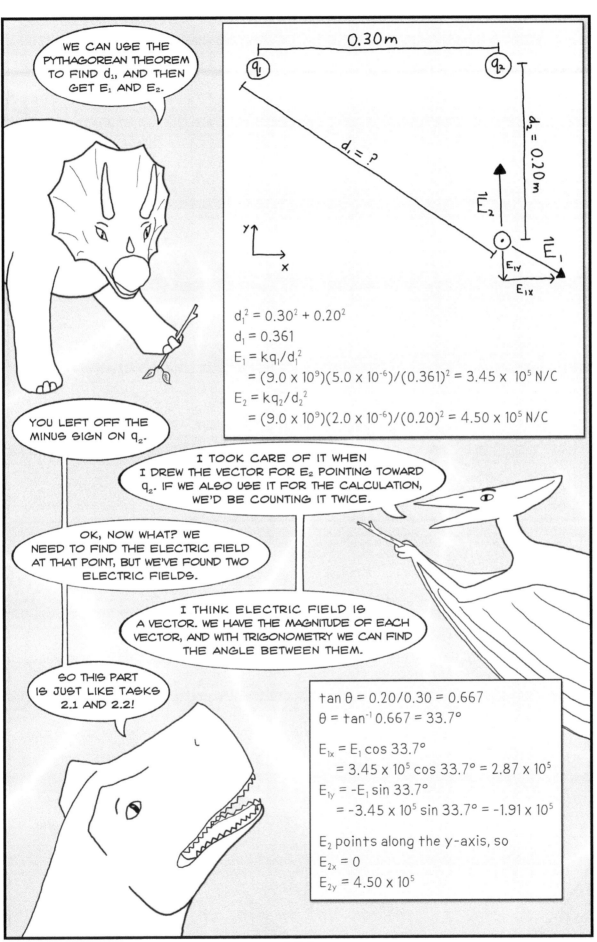

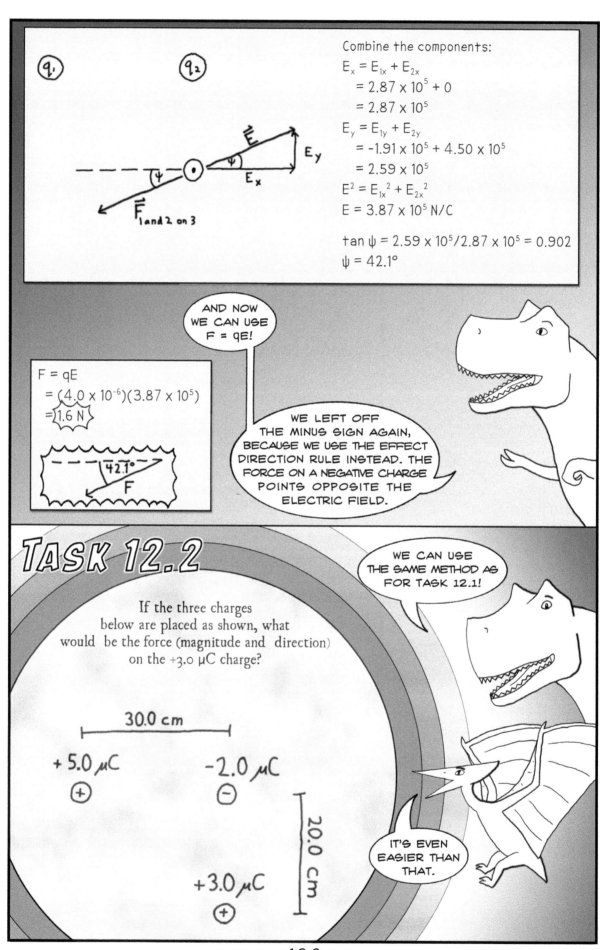

⑨₁ ⑨₂

Combine the components:

$E_x = E_{1x} + E_{2x}$
$\quad = 2.87 \times 10^5 + 0$
$\quad = 2.87 \times 10^5$

$E_y = E_{1y} + E_{2y}$
$\quad = -1.91 \times 10^5 + 4.50 \times 10^5$
$\quad = 2.59 \times 10^5$

$E^2 = E_{1x}^2 + E_{2x}^2$
$E = 3.87 \times 10^5 \text{ N/C}$

$\tan \psi = 2.59 \times 10^5 / 2.87 \times 10^5 = 0.902$
$\psi = 42.1°$

$\vec{E}$
ψ
E_y
E_x
$\vec{F}_{1 \text{ and } 2 \text{ on } 3}$

AND NOW WE CAN USE F = qE!

$F = qE$
$\quad = (4.0 \times 10^{-6})(3.87 \times 10^5)$
$\quad = 1.6 \text{ N}$

42.1°
F

WE LEFT OFF THE MINUS SIGN AGAIN, BECAUSE WE USE THE EFFECT DIRECTION RULE INSTEAD. THE FORCE ON A NEGATIVE CHARGE POINTS OPPOSITE THE ELECTRIC FIELD.

TASK 12.2

WE CAN USE THE SAME METHOD AS FOR TASK 12.1!

If the three charges below are placed as shown, what would be the force (magnitude and direction) on the +3.0 μC charge?

30.0 cm

+ 5.0 μC ⊕

−2.0 μC ⊖

20.0 cm

+3.0 μC ⊕

IT'S EVEN EASIER THAN THAT.

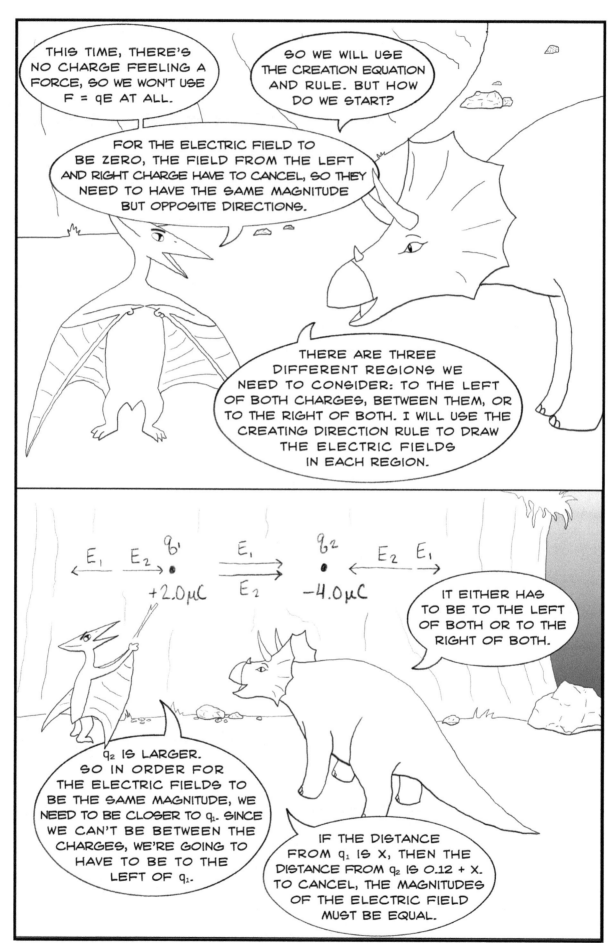

$E_1 = E_2$

$kq_1/x^2 = kq_2/(0.12 + x)^2$

k cancels.

$(2 \times 10^{-6})/x^2 = (4 \times 10^{-6})/(0.12 + x)^2$

Cross-multiply and divide both sides
 by 2×10^{-6}:

$(0.12 + x)^2 = 2x^2$

$0.0144 + 0.24x + x^2 = 2x^2$

$x^2 - 0.24x - 0.0144 = 0$

Solve with the quadratic formula:

x = 0.29 m or -0.05 m.

THE NEGATIVE ANSWER REPRESENTS THE POINT BETWEEN THE TWO CHARGES WHERE THE MAGNITUDES OF THE FIELDS ARE EQUAL, BUT SINCE THEY POINT IN THE SAME DIRECTION THERE, THEY DON'T CANCEL.

THE ONLY ANSWER IS 0.29 METERS TO THE LEFT OF THE +2.0 µC CHARGE.

TASK 12.4

A 7.0 g mass on the end of a string is placed in an electric field of 13,000 N/C, directed from left to right. The system is in equilibrium when the string makes a 23.0° angle to the left of the vertical. What is the charge on the 7.0 g mass?"

THIS IS A STATICS TASK, LIKE 7.9.

1 Choose the 7.0 g mass.

2 Draw a free-body diagram.

qE IS THE ELECTRIC FORCE. BUT HOW DO WE KNOW WHICH WAY IT POINTS?

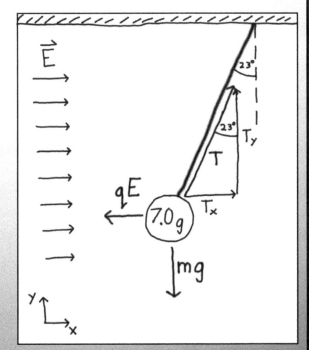

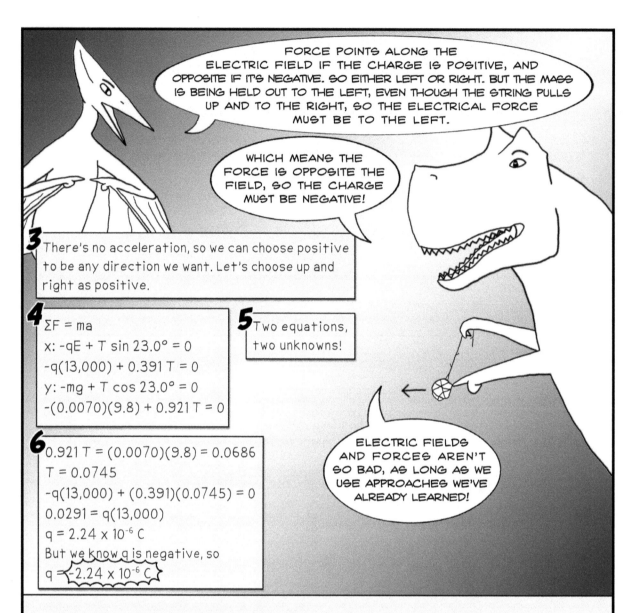

3 There's no acceleration, so we can choose positive to be any direction we want. Let's choose up and right as positive.

4
$\Sigma F = ma$
x: $-qE + T \sin 23.0° = 0$
$-q(13,000) + 0.391\,T = 0$
y: $-mg + T \cos 23.0° = 0$
$-(0.0070)(9.8) + 0.921\,T = 0$

5 Two equations, two unknowns!

6
$0.921\,T = (0.0070)(9.8) = 0.0686$
$T = 0.0745$
$-q(13,000) + (0.391)(0.0745) = 0$
$0.0291 = q(13,000)$
$q = 2.24 \times 10^{-6}\ C$
But we know q is negative, so
$q = -2.24 \times 10^{-6}\ C$

EQUATIONS IN THIS CHAPTER

$$E = kq/d^2 \ (\text{creating}) \qquad F = qE \ (\text{effect})$$

WORDS OF WISDOM

Effect: Force on a positive charge is **along** the electric field; force on a negative charge is **opposite** the electric field

Creating: Electric field created by a positive charge points **away** from that charge; electric field created by a negative charge points **toward** that charge

CHAPTER 13

ELECTRIC POTENTIAL

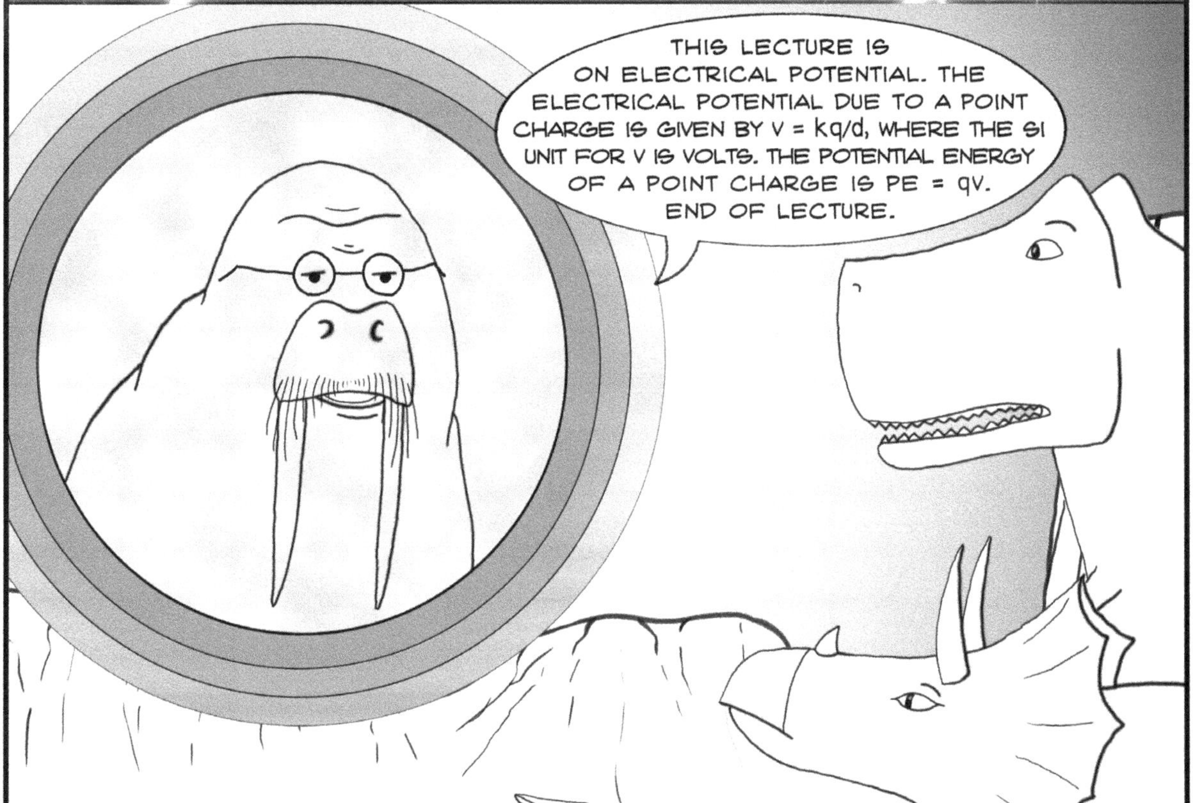

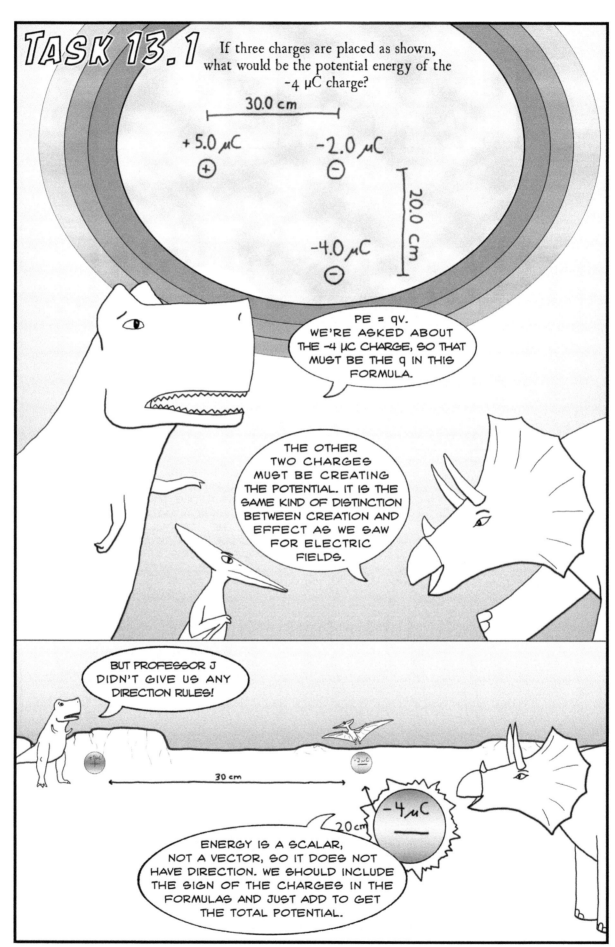

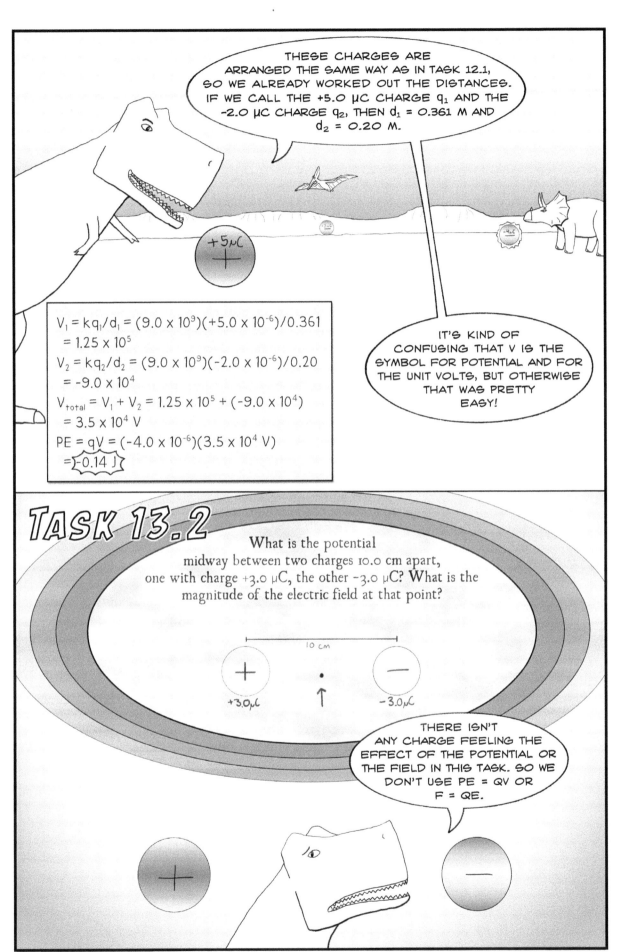

THESE CHARGES ARE ARRANGED THE SAME WAY AS IN TASK 12.1, SO WE ALREADY WORKED OUT THE DISTANCES. IF WE CALL THE +5.0 μC CHARGE q_1 AND THE −2.0 μC CHARGE q_2, THEN $d_1 = 0.361$ M AND $d_2 = 0.20$ M.

+5μC

$V_1 = kq_1/d_1 = (9.0 \times 10^9)(+5.0 \times 10^{-6})/0.361$
  $= 1.25 \times 10^5$

$V_2 = kq_2/d_2 = (9.0 \times 10^9)(-2.0 \times 10^{-6})/0.20$
  $= -9.0 \times 10^4$

$V_{total} = V_1 + V_2 = 1.25 \times 10^5 + (-9.0 \times 10^4)$
  $= 3.5 \times 10^4$ V

$PE = qV = (-4.0 \times 10^{-6})(3.5 \times 10^4$ V$)$
  $= -0.14$ J

IT'S KIND OF CONFUSING THAT V IS THE SYMBOL FOR POTENTIAL AND FOR THE UNIT VOLTS, BUT OTHERWISE THAT WAS PRETTY EASY!

TASK 13.2

What is the potential midway between two charges 10.0 cm apart, one with charge +3.0 μC, the other −3.0 μC? What is the magnitude of the electric field at that point?

10 cm

+3.0μC   −3.0μC

THERE ISN'T ANY CHARGE FEELING THE EFFECT OF THE POTENTIAL OR THE FIELD IN THIS TASK. SO WE DON'T USE PE = QV OR F = QE.

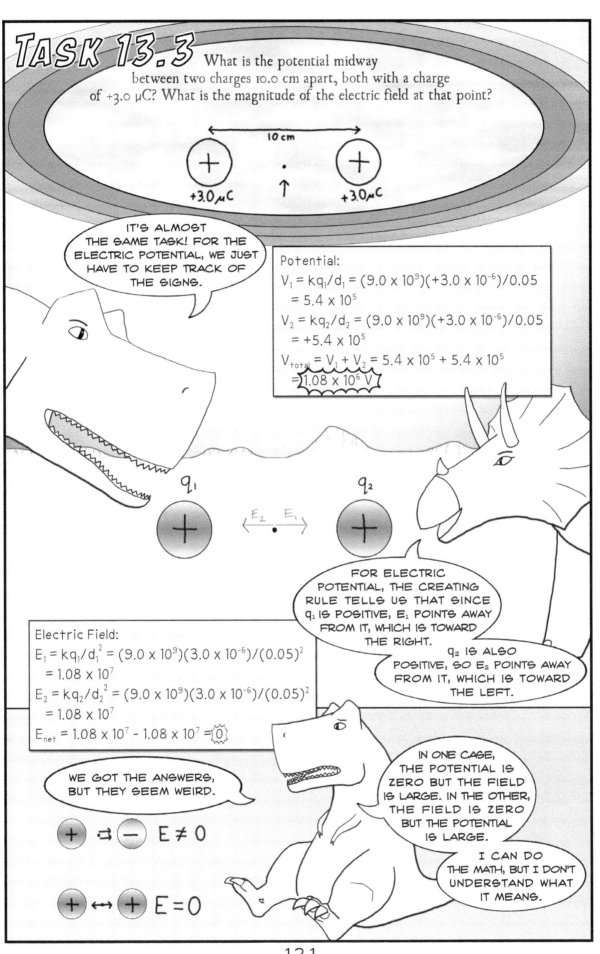

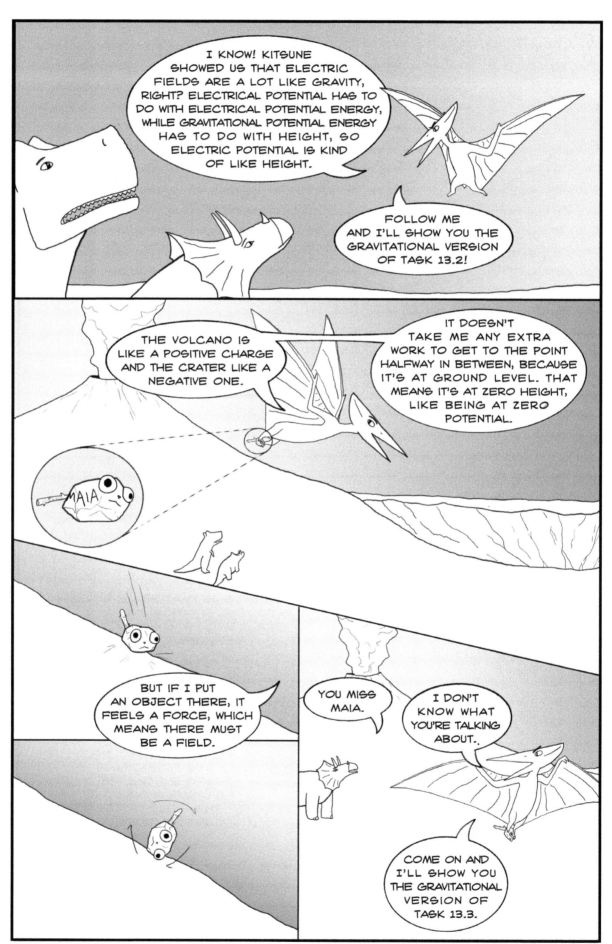

THE "FINAL" STATE IS WHEN THE ALPHA PARTICLE IS 1.0 NANOMETERS FROM THE NUCLEUS.

AT THAT POINT, THE ALPHA PARTICLE IS NOT MOVING FOR AN INSTANT, BECAUSE IT IS AS CLOSE AS IT CAN GET BEFORE THE POSITIVE NUCLEUS THROWS THE POSITIVE ALPHA PARTICLE BACK THE WAY IT CAME.

Final State

Au

1nm

2
There are electrical charges, so there must be electrical potential energy.
$E_f = qV_f$
The charge that gets the potential energy is the alpha particle, so q = +2e. The potential must be due to the gold nucleus, so
$V_f = kq/d = (9.0 \times 10^9)(+79)(1.6 \times 10^{-19})/(1.0 \times 10^{-9}) = 114$
$E_f = qV_f = +2(1.6 \times 10^{-19})(114) = 3.64 \times 10^{-17}$

INITIALLY, THE ALPHA PARTICLE IS MOVING, SO IT HAS KINETIC ENERGY. I THINK WE CAN ASSUME THAT IT STARTS A LONG WAY FROM THE GOLD NUCLEUS, SO IT DOES NOT START WITH MUCH POTENTIAL ENERGY.

3
$E_i = \frac{1}{2} mv^2 = \frac{1}{2} (6.67 \times 10^{-27})v^2 = 3.34 \times 10^{-27} v^2$

4
$W_{nc} = E_f - E_i$
$0 = 3.64 \times 10^{-17} - 3.34 \times 10^{-27} v^2$
$3.34 \times 10^{-27} v^2 = 3.64 \times 10^{-17}$
$v = 1.0 \times 10^5 \text{ m/s}$

WOW! THAT'S FAST. I GUESS THESE PARTICLES ARE SO SMALL IT'S EASY TO GET THEM UP TO HIGH SPEEDS.

YEAH. AND THAT'S WHY WE DIDN'T HAVE TO WORRY ABOUT GRAVITY IN THIS TASK. FOR SUBATOMIC CHARGED PARTICLES, THE ELECTRICAL FORCES ARE WAY LARGER THAN THE GRAVITATIONAL ONES.

TASK 13.5

Two +4.0 μC charges are separated by a distance of 20.0 cm. How much work would it take to move a -3.0 μC charge from a position midway between the two to a position 3.0 cm closer to one of the charges (but still on the line joining them)?

I THINK THE KEY TO THIS TASK IS NOT TO TRY ANY SHORT CUTS. IF WE SOLVE IT STEP BY STEP USING WORK-ENERGY, WE SHOULD BE FINE.

1 Since electrical forces are conservative, the only non-conservative force in this task is whatever is moving the charge. We are asked to find the work, W.

AS IN TASK 5.7, WHETHER WE ASSUME THE INITIAL AND FINAL SPEED ARE ZERO, OR THAT THE SPEED IS CONSTANT, WE GET THE SAME RESULT.

FOR SIMPLICITY, LET US ASSUME THEY ARE ZERO.

2 The "final" state is when q is 13.0 cm from q_1 and 7.0 cm from q_2.

$E_f = qV_f = -3.0 \times 10^{-6} V_f$

The potential is due to the other two charges.

$V_f = kq_1/d_1 + kq_2/d_2$

$= (9.0 \times 10^9)(+4.0 \times 10^{-6})/(0.130)$
$\quad + (9.0 \times 10^9)(+4.0 \times 10^{-6})/(0.070)$

$= 7.91 \times 10^5$

$E_f = qV_f = (-3.0 \times 10^{-6})(7.91 \times 10^5) = -2.37$

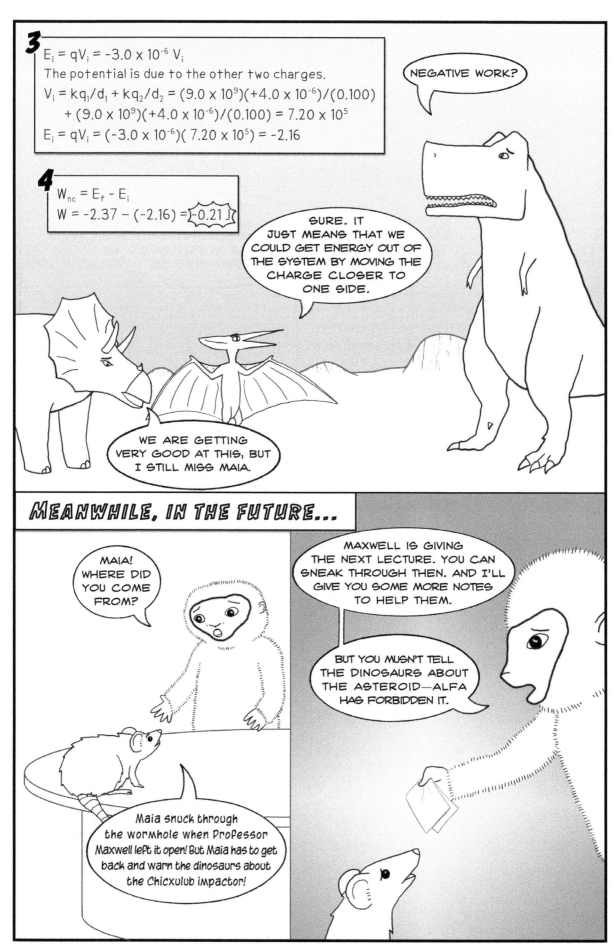

EQUATIONS IN THIS CHAPTER

$V = kq/d$ (creating) $PE = qV$ (effect)

WORDS OF WISDOM

Electric fields are vectors: do not include signs, but draw arrows

Electric potential is a scalar: include signs, but don't draw arrows

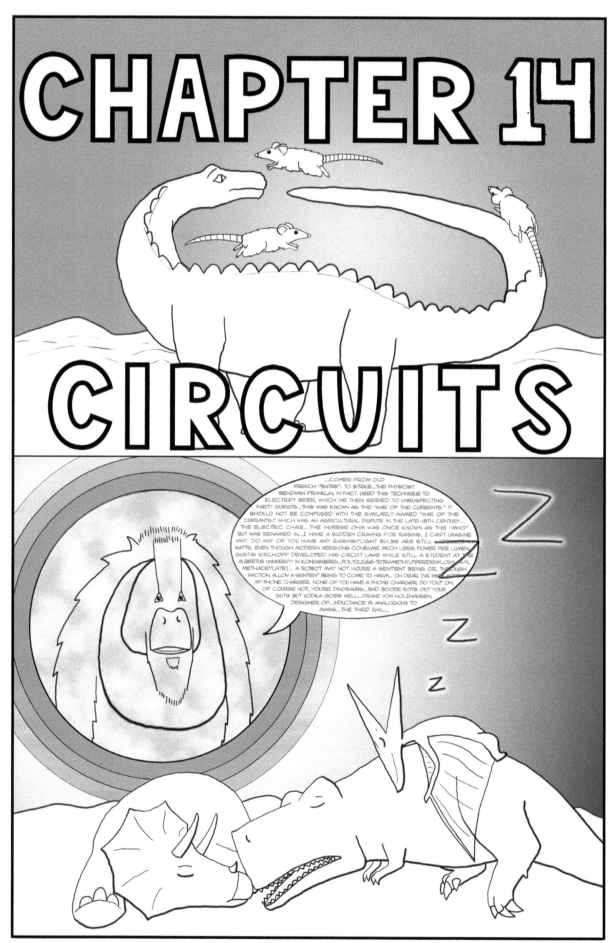

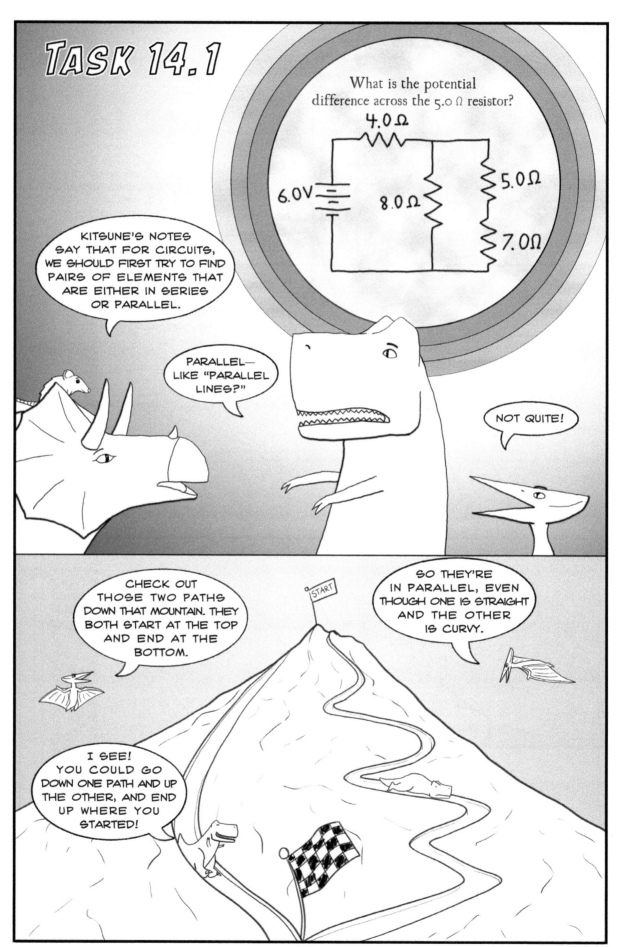

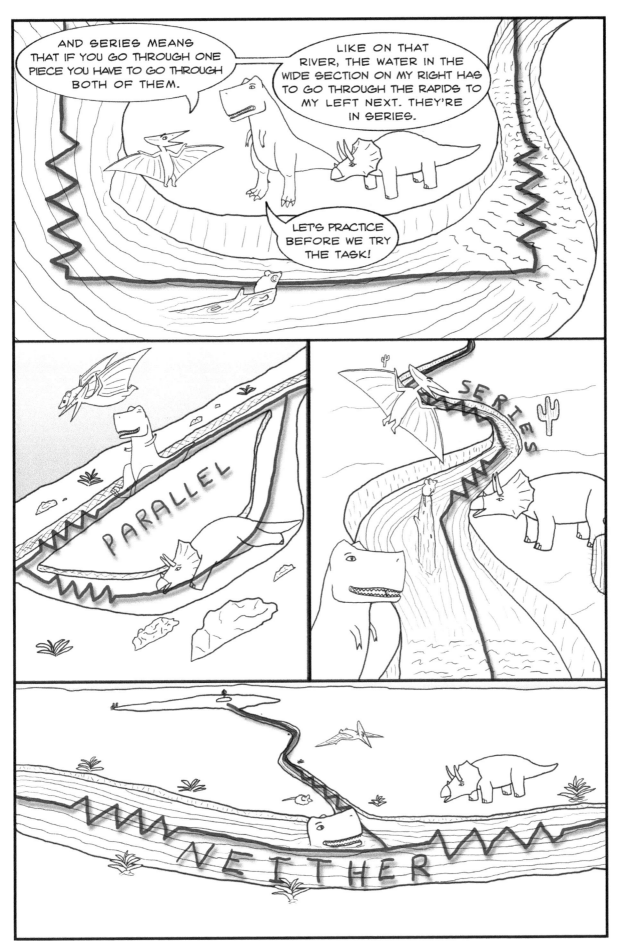

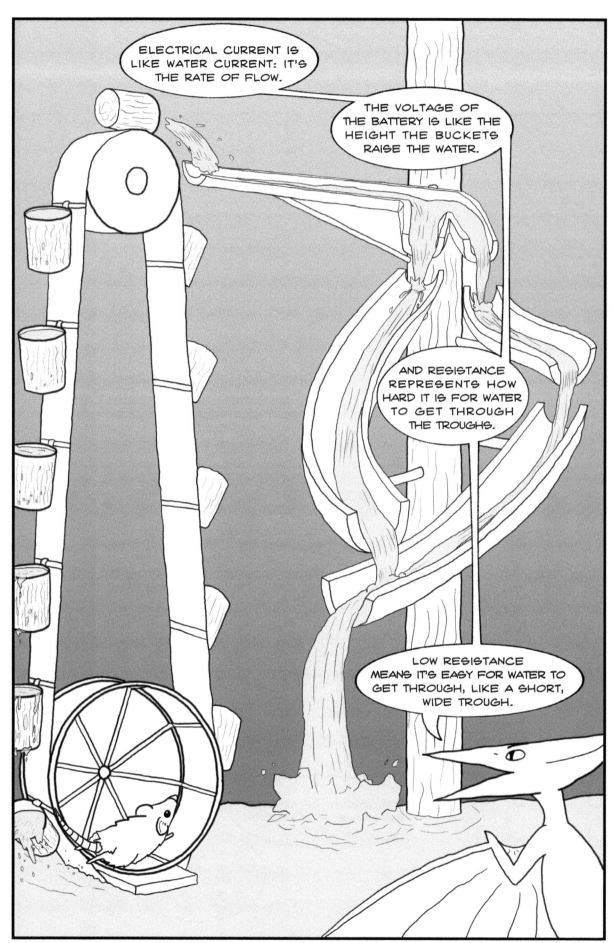

144

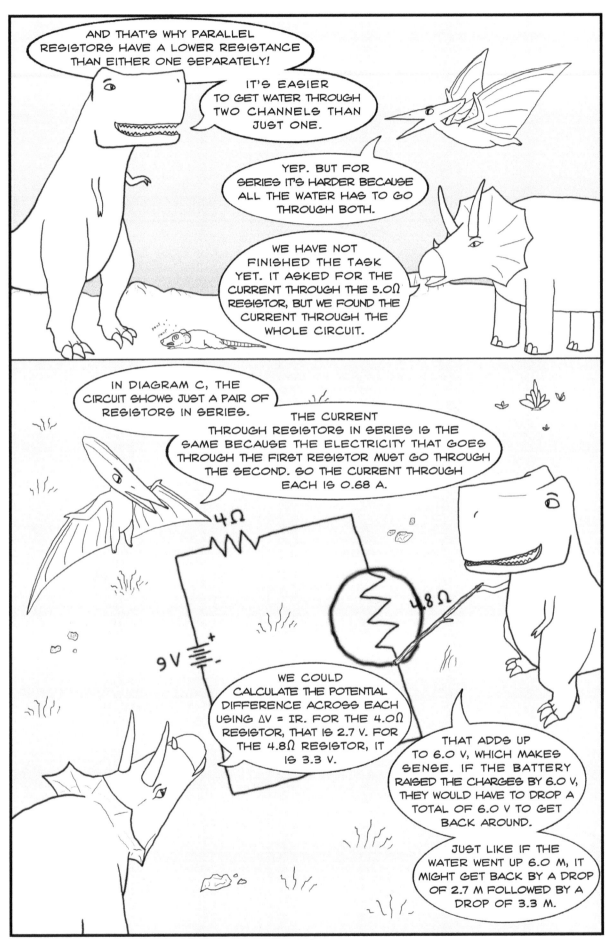

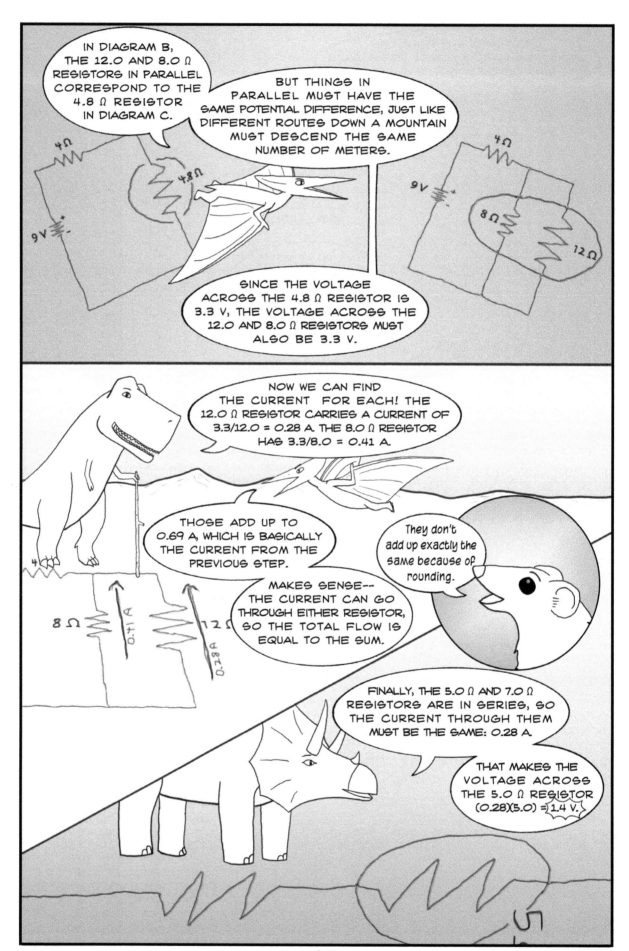

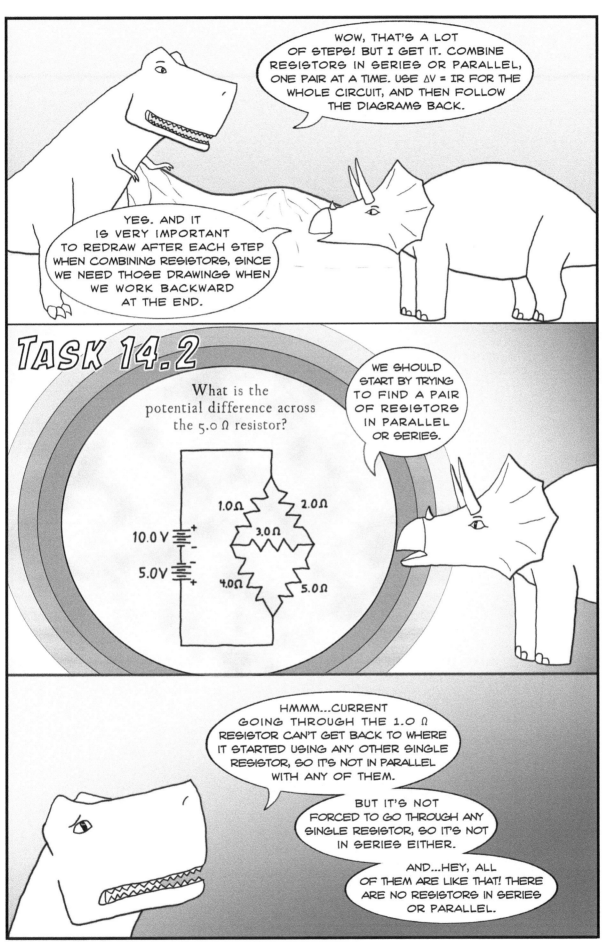

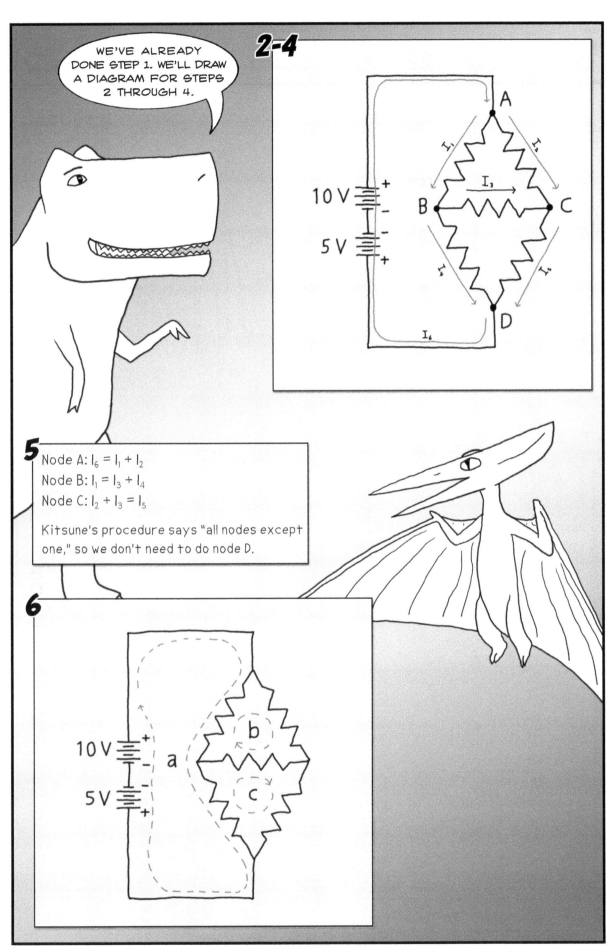

7

Path a: $-1.0\,I_1 - 4.0\,I_4 - 5.0 + 10.0 = 0$

Path b: $-2.0\,I_2 + 3.0\,I_3 + 1.0\,I_1 = 0$

Path c: $-3.0\,I_3 - 5.0\,I_5 + 4.0\,I_4 = 0$

8

3 node equations + 3 path equations = 6 equations

6 unknown currents

As many equations as unknowns!

9

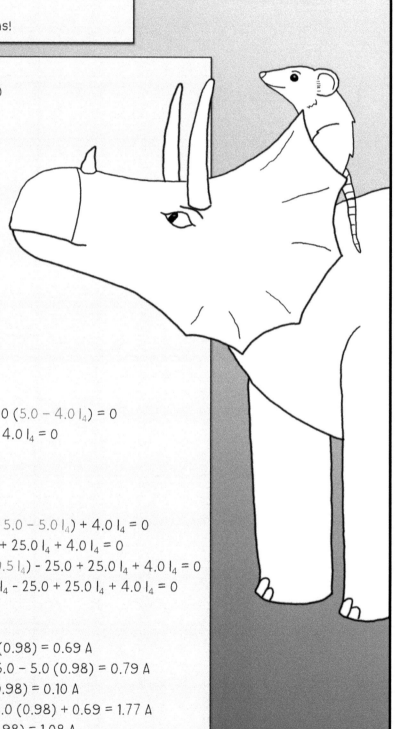

Path a:

$-1.0\,I_1 - 4.0\,I_4 - 5.0 + 10.0 = 0$

$I_1 = 5.0 - 4.0\,I_4$

Node A:

$I_6 = I_1 + I_2$

$I_6 = 5.0 - 4.0\,I_4 + I_2$

Node B:

$I_1 = I_3 + I_4$

$5.0 - 4.0\,I_4 = I_3 + I_4$

$I_3 = 5.0 - 5.0\,I_4$

Node C:

$I_2 + I_3 = I_5$

$I_5 = I_2 + 5.0 - 5.0\,I_4$

Path b:

$-2.0\,I_2 + 3.0\,I_3 + 1.0\,I_1 = 0$

$-2.0\,I_2 + 3.0\,(5.0 - 5.0\,I_4) + 1.0\,(5.0 - 4.0\,I_4) = 0$

$-2.0\,I_2 + 15.0 - 15.0\,I_4 + 5.0 - 4.0\,I_4 = 0$

$2.0\,I_2 = 20.0 - 19.0\,I_4$

Path c:

$-3.0\,I_3 - 5.0\,I_5 + 4.0\,I_4 = 0$

$-3.0\,(5.0 - 5.0\,I_4) - 5.0\,(I_2 + 5.0 - 5.0\,I_4) + 4.0\,I_4 = 0$

$-15.0 + 15.0\,I_4 - 5.0\,I_2 - 25.0 + 25.0\,I_4 + 4.0\,I_4 = 0$

$-15.0 + 15.0\,I_4 - 5.0\,(10.0 - 9.5\,I_4) - 25.0 + 25.0\,I_4 + 4.0\,I_4 = 0$

$-15.0 + 15.0\,I_4 - 50.0 + 47.5\,I_4 - 25.0 + 25.0\,I_4 + 4.0\,I_4 = 0$

$91.5\,I_4 = 90.0$

$I_4 = 0.98\ \text{A}$

$I_2 = 10.0 - 9.5\,I_4 = 10.0 - 9.5\,(0.98) = 0.69\ \text{A}$

$I_5 = I_2 + 5.0 - 5.0\,I_4 = 0.69 + 5.0 - 5.0\,(0.98) = 0.79\ \text{A}$

$I_3 = 5.0 - 5.0\,I_4 = 5.0 - 5.0\,(0.98) = 0.10\ \text{A}$

$I_6 = 5.0 - 4.0\,I4 + I_2 = 5.0 - 4.0\,(0.98) + 0.69 = 1.77\ \text{A}$

$I_1 = 5.0 - 4.0\,I_4 = 5.0 - 4.0(0.98) = 1.08\ \text{A}$

$V_5 = I_5\,(5) = (0.79)(5) = 4.0\ \text{V}$

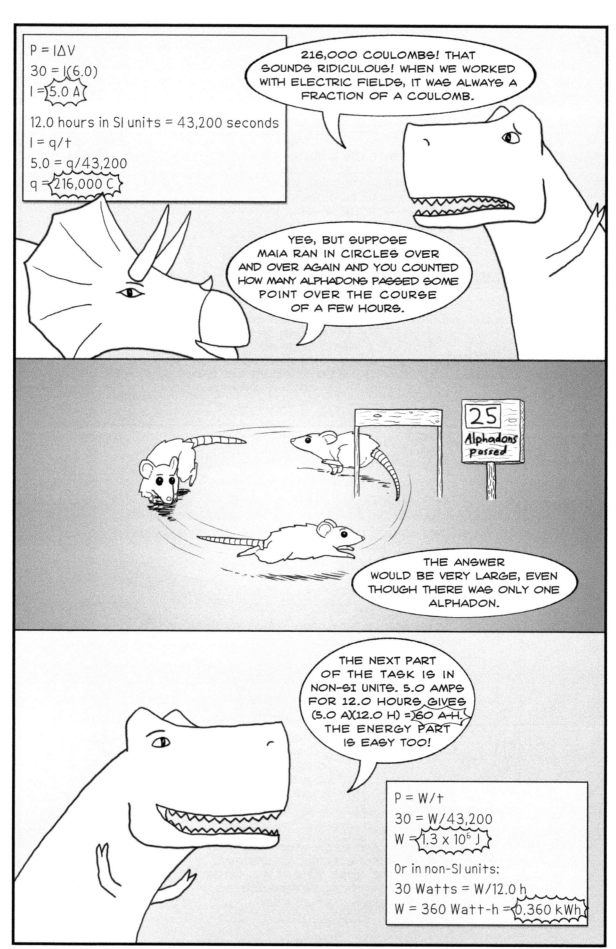

TASK 14.4

A 1.5 V battery has an internal resistance of 2.0 Ω. If it is hooked up to a 1.0 Ω light bulb, how much power actually reaches the bulb? What is the "terminal potential difference" of the battery?

The resistors are in series. They can be replaced by a single resistor of 1.0 + 2.0 = 3.0 Ω.

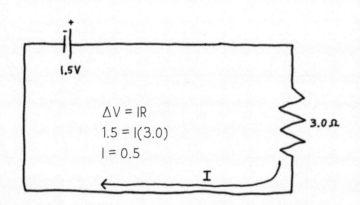 

$$\Delta V = IR$$
$$1.5 = I(3.0)$$
$$I = 0.5$$

Now go back to the first diagram. Since the resistors are in series, the current is the same. The voltage across the internal resistance is $\Delta V = IR = (0.5)(2.0) = 1.0$

The voltage across the light bulb is
$$\Delta V = IR = (0.5)(1.0) = 0.5$$

The power for the light bulb is
$$P = I\Delta V = (0.5)(0.5) = 0.25 \text{ W.}$$

There are two ways to get the terminal potential across the battery:

1. The battery is responsible for the potential difference across the light bulb, so it must be 0.5 V.

2. the battery has an emf of 1.5 V, 1.0 V of which is used up by the internal resistance, leaving 0.5 V.

TASK 14.5

A 3.0 Ω wire is cut into three pieces of equal length. The three pieces are then bound together, side by side, to create, in effect, a new wire. What is the resistance of the new wire?

THIS SOUNDS LIKE THE PROPORTIONAL REASONING WE DID FOR TASK 4.5. BUT WE NEED AN EQUATION THAT TELLS US HOW RESISTANCE DEPENDS ON LENGTH.

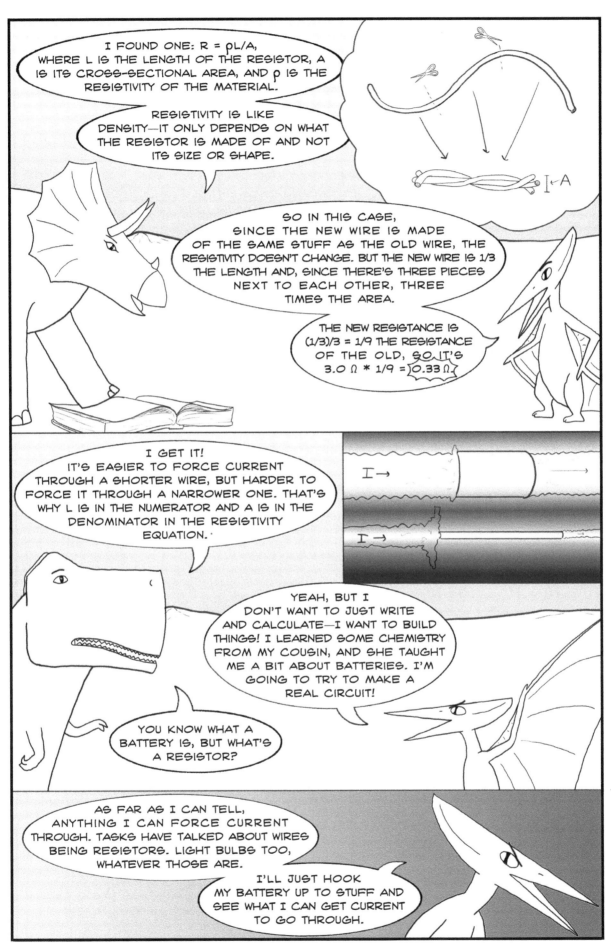

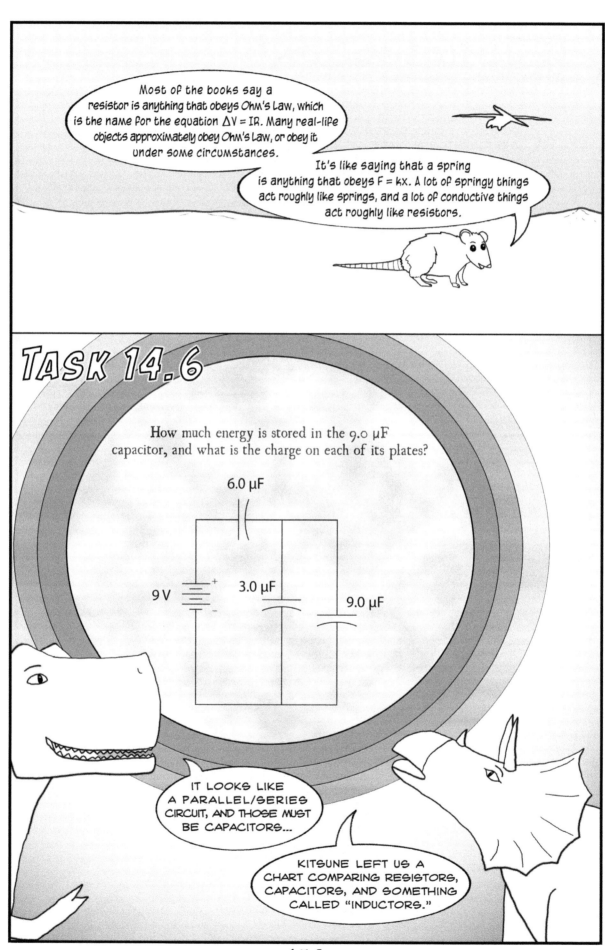

The 12.0 and 6.0 μF capacitors are in series, so
$1/C = 1/(1.20 \times 10^{-5}) + 1/(6.0 \times 10^{-6})$
$C = 4.0 \times 10^{-6}$

9 V

4.0 μF

$Q = C\Delta V$
$Q = (4.0 \times 10^{-6})(9.0) = 3.6 \times 10^{-5}$

WE GOT THE OVERALL CHARGE IN THE CIRCUIT, SO NOW WE HAVE TO WORK OUR WAY BACK THROUGH THE EARLIER DIAGRAMS, LIKE WE DID FOR RESISTORS.

BUT CAPACITORS IN SERIES COMBINE LIKE RESISTORS IN PARALLEL, AND VICE-VERSA. DOES THAT MEAN THE VOLTAGE AND CURRENT RULES ARE SWITCHED TOO?

NO. VOLTAGE IS ALWAYS THE SAME IN PARALLEL, AND ALWAYS ADDS IN SERIES.

CURRENT IS JUST CHARGE IN MOTION, SO CHARGE AND CURRENT ARE ALWAYS THE SAME IN SERIES AND ADD IN PARALLEL. IT DOESN'T MATTER WHETHER WE'VE GOT RESISTORS OR CAPACITORS.

I HAVE COME UP WITH A MNEMONIC FOR HOW RESISTANCES AND CAPACITANCES COMBINE.

THE SYMBOLS FOR TWO RESISTORS IN A ROW LOOK LIKE ONE BIG RESISTOR, SO THEY ADD IN SERIES. (THE SAME IS TRUE FOR INDUCTORS.)

BUT TWO CAPACITORS IN PARALLEL LOOK LIKE ONE BIG CAPACITOR, SO THEY ADD IN PARALLEL.

I DON'T THINK THAT'S JUST A MNEMONIC--THE SYMBOLS LOOK LIKE THE REAL THINGS, AND REFLECT HOW THEY BEHAVE!

BACK TO THE TASK!

C STANDS FOR CAPACITANCE BUT ALSO FOR COULOMBS.

I GUESS PHYSICISTS WERE RUNNING OUT OF LETTERS!

The 12.0 and 6.0 µF capacitors are in series, so each has charge 3.6×10^{-5} C.

$Q = C\Delta V$

$3.6 \times 10^{-5} = (12.0 \times 10^{-6})(\Delta V) = 3.0$

$3.6 \times 10^{-5} = (6.0 \times 10^{-6})(\Delta V) = 6.0$

They add up to 9.0 V, as they should.

The 12.0 µF capacitor consists of 3.0 and 9.0 µF capacitors in parallel, so each has voltage of 3.0 V.

$Q = C\Delta V$

$Q = (3.0 \times 10^{-6})(3.0) = 9.0 \times 10^{-6}$ C

$Q = (9.0 \times 10^{-6})(3.0) = 2.7 \times 10^{-5}$ C

They add up to 3.6×10^{-5} C, as they should.

Energy on 9.0 µF capacitor:

$\frac{1}{2} C(\Delta V)^2 = \frac{1}{2}(9.0 \times 10^{-6})(3.0)^2 = 4.0 \times 10^{-5}$ J

TASK 14.7

30 ms after the battery was disconnected, what is the charge on the capacitor and the current through the circuit? How long will it take for the voltage across the capacitor to reach 8.0 V? Assume the battery had been connected for a long time before the switch was thrown.

30 µF

9 V

1500 Ω

THIS CIRCUIT HAS BOTH A RESISTOR AND A CAPACITOR IN IT.

KITSUNE GAVE US NOTES ABOUT THIS, BUT THEY ARE NOT AS CLEAR AS USUAL.

RC Circuit: $\tau = RC$

LR Circuit: $\tau = L/R$

Start at max, drop to zero: multiply by $e^{-t/\tau}$

Start at zero, rise to max: multiply by $(1 - e^{-t/\tau})$

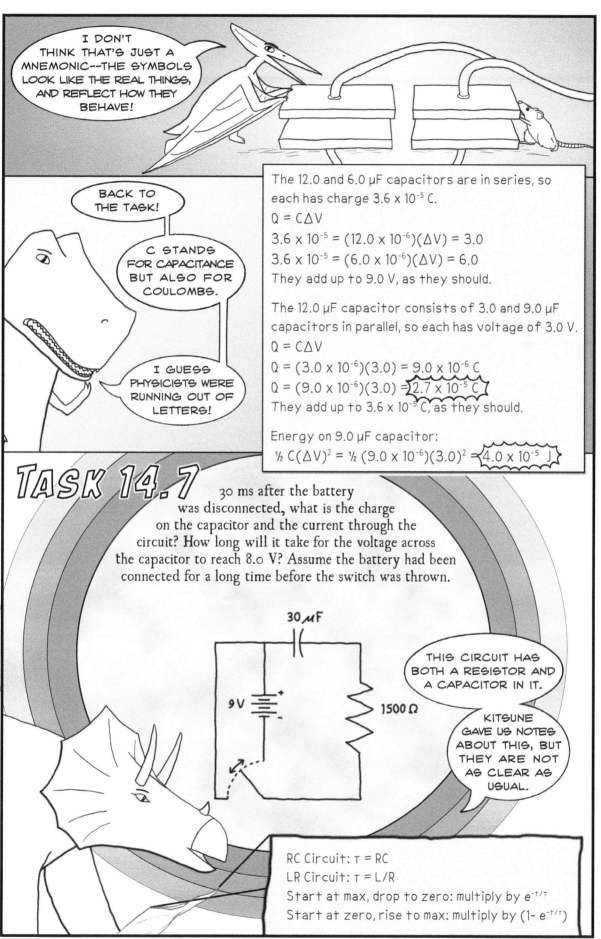

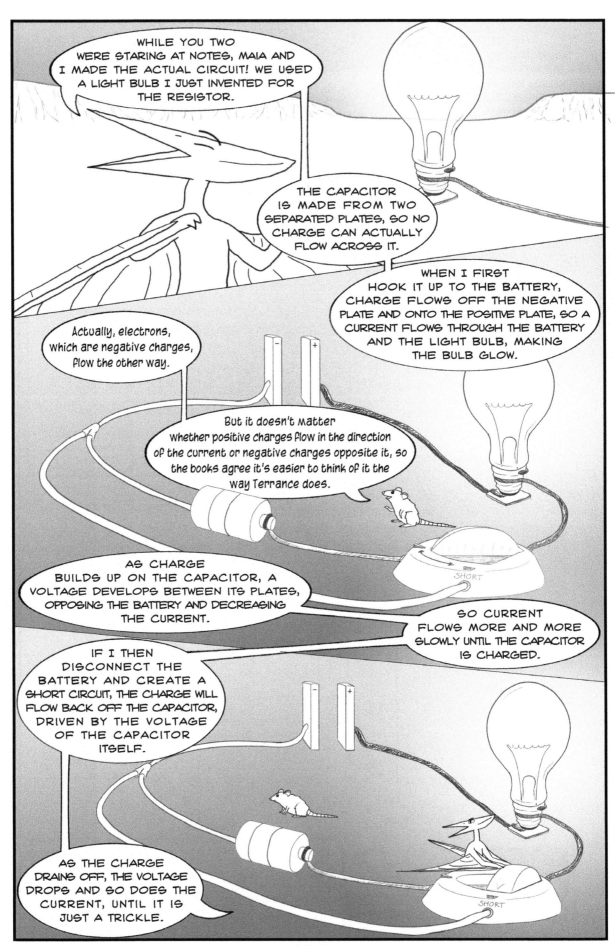

160

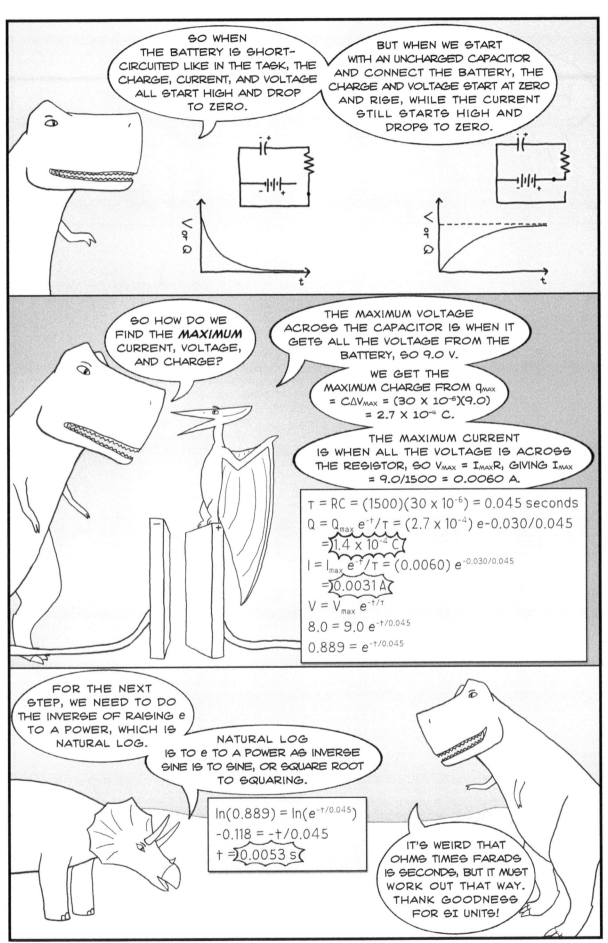

KITSUNE'S CIRCUITS CHART

	Resistors	Capacitors	Inductors
Units	Ohms (Ω)	Farads (F)	Henrys(H)
In series	Add	$1/C = 1/C_1 + 1/C_2$	Add
In parallel	$1/R = 1/R_1 + 1/R_2$	Add	$1/L = 1/L_1 + 1/L_2$
Voltage	$\Delta V = IR$	$Q = C\Delta V$	$\Delta V = -L\Delta I/\Delta t$
Geometry	$R = pL/A$	$Q = \kappa\epsilon_0 A/d$ (parallel-plate)	$L = \mu AN^2/\text{length}$ (solenoid)

TIME-DEPENDENT CIRCUITS

RC Circuit: $\tau = RC$

LR Circuit: $\tau = L/R$

Start at max, drop to zero: multiply by $e^{-t/\tau}$

Start at zero, rise to max: multiply by $(1 - e^{-t/\tau})$

WORDS OF WISDOM

Series: If you go through one, you must go through the other

Parallel: If you go through one and then the other, you end up where you started

When solving using parallel/series, draw each picture! If you ever find yourself using information from more than one picture back, you are making a mistake

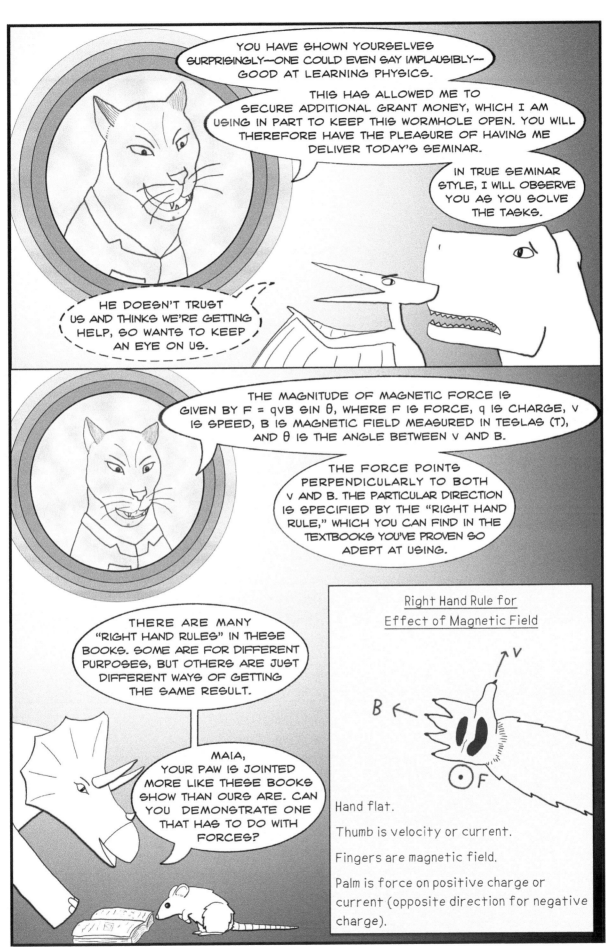

TASK 15.1

In each case, identify the direction of the missing vector (v, B, or F). Assume all charges are positive and that the magnetic field, if not known, is perpendicular to the velocity. Dots represent a vector pointing out of the page, and x's into the page.

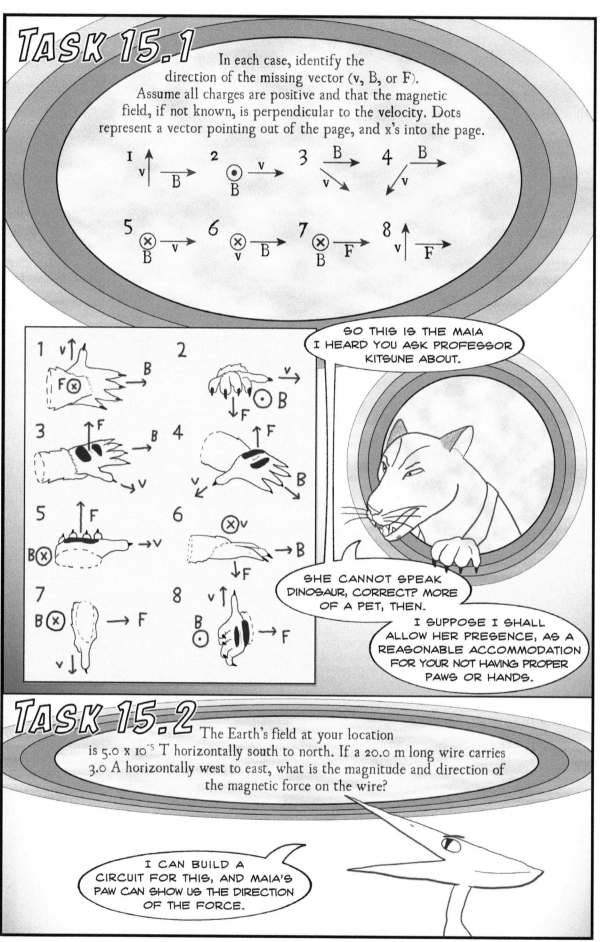

1 v ↑   B →
2 ⊙ B   v →
3 B →   v ↘
4 B →   v ↙
5 ⊗ B   v →
6 ⊗ v   B →
7 ⊗ B   F →
8 v ↑   F →

SO THIS IS THE MAIA I HEARD YOU ASK PROFESSOR KITSUNE ABOUT.

SHE CANNOT SPEAK DINOSAUR, CORRECT? MORE OF A PET, THEN.

I SUPPOSE I SHALL ALLOW HER PRESENCE, AS A REASONABLE ACCOMMODATION FOR YOUR NOT HAVING PROPER PAWS OR HANDS.

TASK 15.2

The Earth's field at your location is 5.0×10^{-5} T horizontally south to north. If a 20.0 m long wire carries 3.0 A horizontally west to east, what is the magnitude and direction of the magnetic force on the wire?

I CAN BUILD A CIRCUIT FOR THIS, AND MAIA'S PAW CAN SHOW US THE DIRECTION OF THE FORCE.

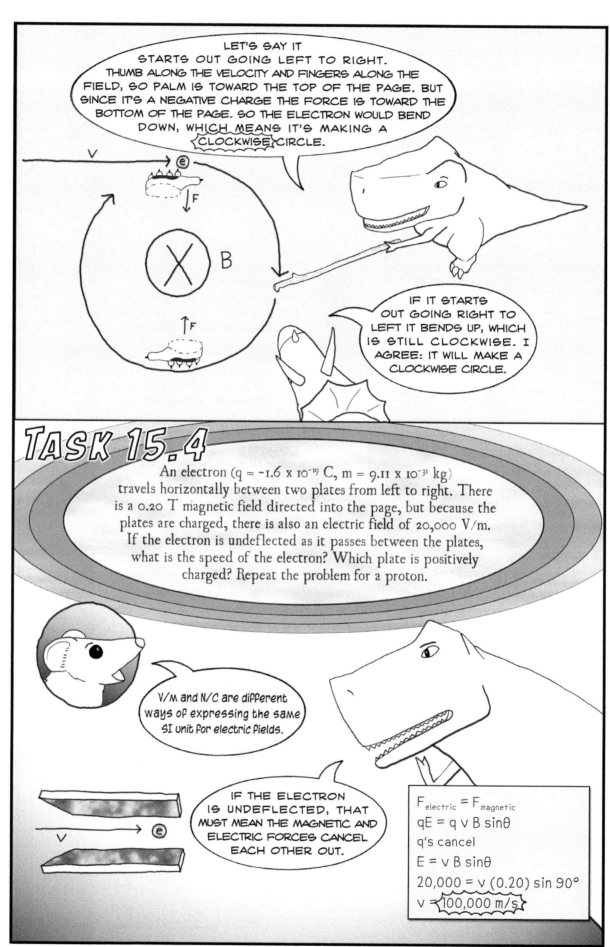

TASK 15.4

An electron (q = -1.6 x 10⁻¹⁹ C, m = 9.11 x 10⁻³¹ kg) travels horizontally between two plates from left to right. There is a 0.20 T magnetic field directed into the page, but because the plates are charged, there is also an electric field of 20,000 V/m. If the electron is undeflected as it passes between the plates, what is the speed of the electron? Which plate is positively charged? Repeat the problem for a proton.

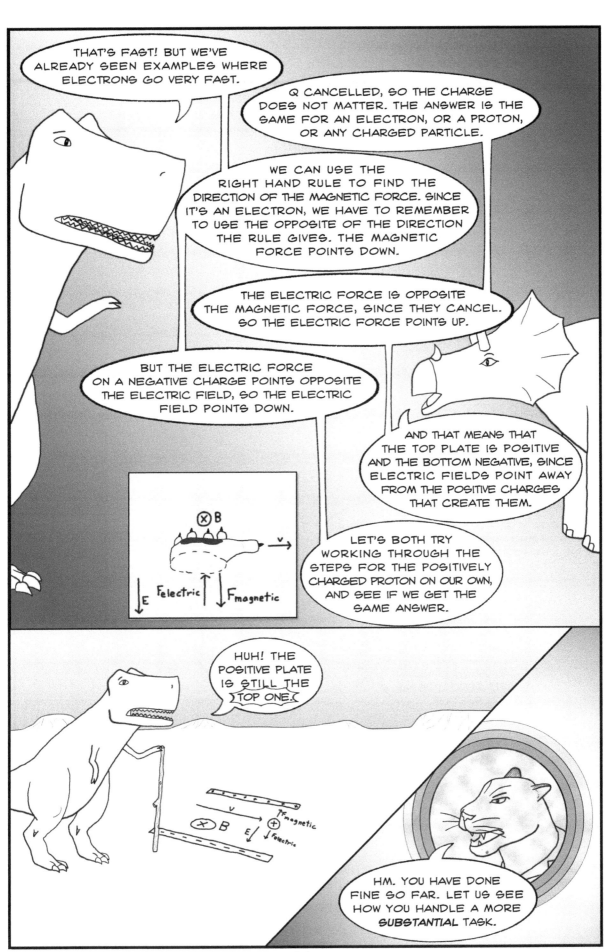

TASK 15.5

An alpha particle
$(q = +2e = 3.2 \times 10^{-19}$ C, $m = 6.67 \times 10^{-27}$ kg$)$
is accelerated from rest by a 200 V potential difference
so that it travels in the plane of the page. It then enters a region
where there is a 0.2 T magnetic field directed perpendicular to the
plan of the page. It is found to circle clockwise.

a) What is the radius of the resulting circular path?
b) Does the magnetic field point into or out of the page?

When it reaches the bottom of its circle, charge is applied to the two
plates shown (one is given a positive charge and the other a negative
charge of the same magnitude). Each plate has an area of 50 cm².

c) What is the magnitude of the charge that must be applied to each
plate to cause the alpha particle to travel undeflected?
d) Which plate is positively charged?

IT'S LONG, BUT IF WE TAKE IT A PIECE AT A TIME WE'LL BE FINE. THE FIRST PART SOUNDS LIKE TASK 15.3.

IT'S TEMPTING TO PUT 200 V IN FOR V, BUT THE V IN THIS FORMULA IS SPEED, AND 200 V IS POTENTIAL DIFFERENCE ("VOLTAGE").

$F = ma$
$q v B \sin\theta = mv^2/r$
$q B \sin\theta = m v/r$
$(3.2 \times 10^{-19})(0.2) \sin 90°$
$= (6.67 \times 10^{-27})v/r$

OH! BUT I CAN USE VOLTAGE TO GET SPEED, USING ENERGY.

1 Since electrical forces are conservative, there are no non-conservative forces in this task.

2-3 The "final" state is when the particle has been accelerated, so it has a final kinetic energy, $\frac{1}{2} mv^2 = \frac{1}{2} (6.67 \times 10^{-27}) v^2 = (3.34 \times 10^{-27}) v^2$. It starts at rest, so there is no initial kinetic energy.
Since PE = qV,
$\Delta PE = q\Delta V = (3.2 \times 10^{-19}) (200) = 6.4 \times 10^{-17}$

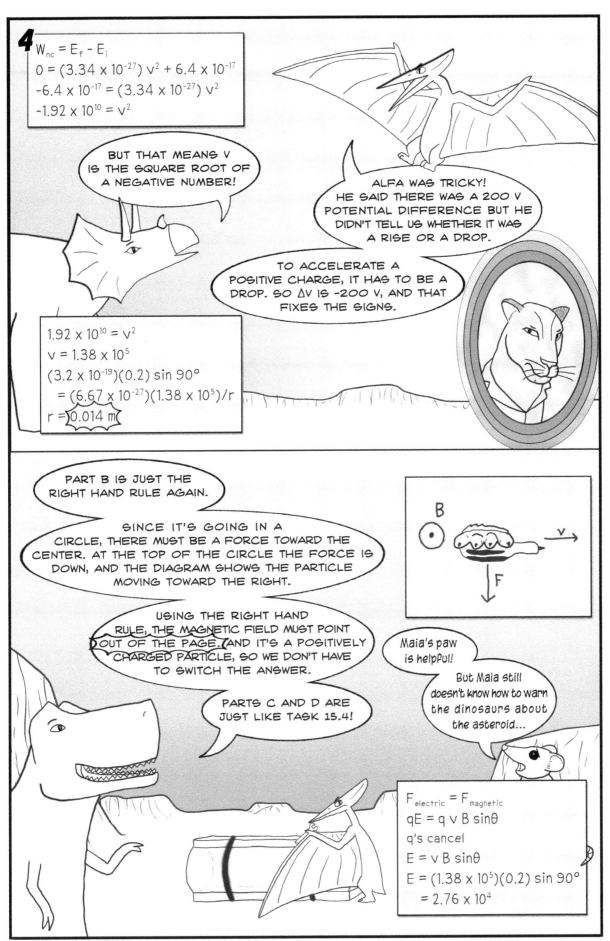

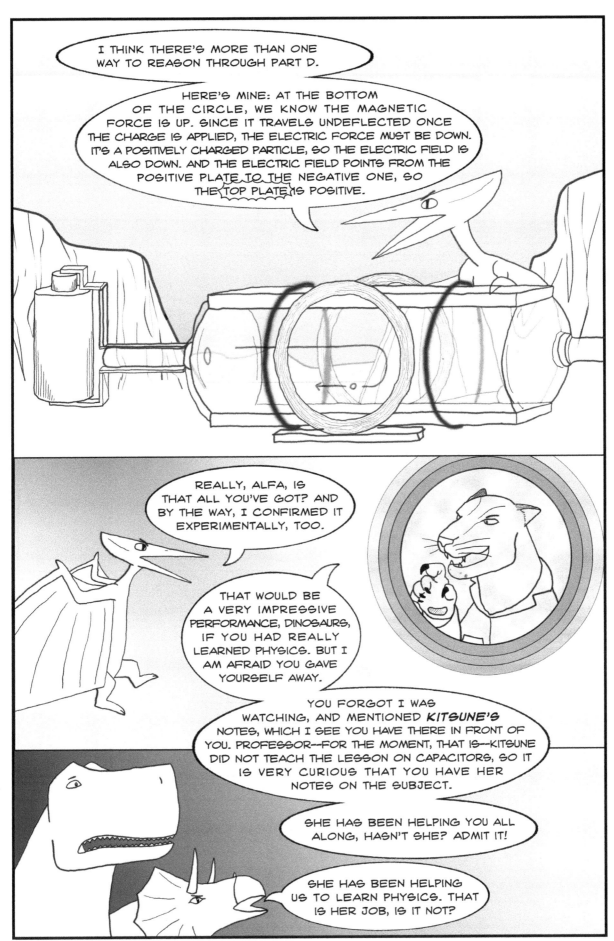

HUH. A STRAIGHT WIRE PRODUCES A MAGNETIC FIELD THAT CIRCLES!

TASK 15.7

a) A current points into the page. Does the magnetic field created by this current circle clockwise or counterclockwise?

b) A +2.0 μC charge with mass 25 grams is travelling left to right on the page with a speed of 2.0 x 10⁵ m/s. It then enters a region where there is a 0.050 T magnetic field pointing out of the page. Does the charge circle clockwise or counterclockwise? What is the radius of curvature of its path?

c) A 3.0 A current flows counterclockwise in a 12-turn circular loop of radius 1.0 cm. What is the magnitude and direction of the magnetic field created in the center of the loop?

A IS JUST LIKE TASK 15.6. IF WE USE THE "CREATING RIGHT HAND RULE", CHOOSE ANY POINT NEAR THE WIRE, AND REMEMBER THAT THE PATTERN IS CIRCULAR, IT'S CLEARLY CLOCKWISE.

PART B IS ALSO STRAIGHTFORWARD, SINCE IT IS LIKE TASK 15.3. USING THE EFFECT RIGHT HAND RULE WITH A POSITIVE CHARGE, THE MAGNETIC FORCE INITIALLY POINTS DOWNWARD, SO IT CIRCLES CLOCKWISE.

B

$F = ma$

$q v B \sin\theta = mv^2/r$

$q B \sin\theta = m v/r$

$(2.0 \times 10^{-6})(0.050)\sin 90°$

$\quad = (0.025)(2.0 \times 10^5)/r$

$r = 5.0 \times 10^{10}$ m

THAT'S HUGE--WAY BIGGER THAN THE ENTIRE EARTH! YOU MUST HAVE MADE A MISTAKE.

I HAVE CHECKED IT. I DO NOT THINK THERE IS A MISTAKE.

HMM...THAT'S A TYPICAL CHARGE IN A TYPICAL MAGNETIC FIELD, WITH THE MASS OF A PEBBLE AND A VERY HIGH SPEED, AND IT'S STILL ALMOST UNDEFLECTED BY THE MAGNETIC FIELD. THAT EXPLAINS WHY WE DON'T USUALLY NOTICE MAGNETIC EFFECTS MUCH WITH ORDINARY OBJECTS--ONLY TINY PARTICLES LIKE ELECTRONS.

The textbooks also mention things like "kitchen magnets." Apparently the way they work is related to the fundamentals of magnetism we have been learning, but much more complicated. Still, they can be thought of as depending on interactions between individual electrons moving within the materials, so Terrance is not wrong.

WHAT ABOUT PART C? WE NEED THE MAGNETIC FIELD CREATED BY A CIRCULAR LOOP, BUT OUR EQUATION IS FOR THE MAGNETIC FIELD CREATED BY A LONG, STRAIGHT WIRE.

I'M LEARNING TO BE CAREFUL ABOUT NOT USING AN EQUATION JUST BECAUSE IT HAS THE RIGHT VARIABLES IN IT!

THE BOOKS SAY THAT IN THE CENTER OF A CIRCULAR LOOP THE MAGNETIC FIELD IS $B = \mu_0 nI/2R$, WHERE n IS THE NUMBER OF TURNS OF WIRE THAT MAKE UP THE LOOP. INSIDE A **SOLENOID** (A COIL OF WIRE), $B = \mu_0 nI$, WHERE n IS THE NUMBER OF TURNS OF THE COIL PER UNIT LENGTH.

SOLUTION 15.7 (CONTINUED)

C

Center of circular loop:

$B = \mu_0 I/2R$

$= (4\pi \times 10^{-7})(12)(3.0)/(2)(0.01)$

$= 0.0023$ T

Solenoid (unrelated to Task 15.7)

$L = 0.36 m$

The solenoid above has 18 turns. $n = \frac{N}{L} = \frac{18}{0.36} = 50$ turns/m

TASK 15.8

Two parallel wires, 8.0 cm apart, carry current in the same direction. The first wire carries 2.0 A, the second 3.0 A. What is the force per unit length between them? What direction does it point?

WE CAN FIND THE DIRECTION LIKE THAT TOO. I'LL MAKE A SKETCH WITH THE 2.0 A WIRE ON THE LEFT AND THE 3.0 A ON THE RIGHT, WITH BOTH CURRENTS POINTING UP.

USING THE CREATING RIGHT HAND RULE, THE FIELD DUE TO THE 2.0 A WIRE AT THE LOCATION OF THE 3.0 A WIRE POINTS INTO THE PAGE.

THEN, USING THE EFFECT RULE, THE FORCE ON THE 3.0 A WIRE DUE TO A FIELD POINTING INTO THE PAGE IS DIRECTED TO THE LEFT. SO THE WIRES MUST ATTRACT EACH OTHER.

2A   3A

8cm

Simplicio arbitrarily chose the direction to draw the currents in, but direction is just a matter of perspective. Turn the page upside down, and currents that went up now point down. Look from behind and the wire that was on the left is on the right. But attraction is always attraction, from any angle.

$B = \mu_o I / 2\pi r$
$= (4\pi \times 10^{-7})(2.0)/[2\pi (0.08)]$
$= 5.0 \times 10^{-6}$

$F = I L B \sin\theta$
$= (3.0) L (5 \times 10^{-6}) \sin 90°$
$= 1.5 \times 10^{-5} L$

$F/L = 1.5 \times 10^{-5}$ N/m

AGAIN, A VERY SMALL FORCE, DESPITE HAVING NORMAL-SIZED CURRENTS SEPARATED BY A TYPICAL REAL-WORLD DISTANCE. MAGNETIC FORCES ON ORDINARY OBJECTS TEND TO BE VERY SMALL!

I MUST ADMIT, DINOSAURS, THAT YOU HAVE SHOWN SOME SKILL. IT IS A PITY THAT YOU ARE ALMOST OUT OF TIME.

!

EQUATIONS IN THIS CHAPTER

$$F = qvB\sin(\theta) \text{ (point charge)}$$

$$F = ILB\sin(\theta) \text{ (segment of wire)}$$

$$B = \mu_o I / 2\pi r \text{ (long, straight wire)}$$

$$B = \mu_o NI / 2R \text{ (center of circular loop)}$$

$$B = \mu_o nI = B = \mu_o NI / L \text{ (inside a solenoid)}$$

RIGHT HAND RULE (EFFECT)

Hand flat.
Thumb is velocity or current.
Fingers are magnetic field.
Palm is force on positive charge or current
(opposite direction for negative charge).

RIGHT HAND RULE (CREATING)

Hand flat.
Thumb along current.
Fingers from wire to point of interest.
Palm is magnetic field.

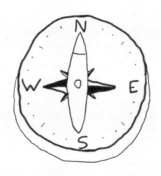

TASK 16.1

A 10.0 cm diameter, 100-turn loop is aligned so that its plane is perpendicular to a 0.20 T magnetic field. If that field is reduced to zero in 50 ms, what is the induced EMF? If the loop has 2.0 Ω resistance, how much current flows? How much power is dissipated by the resistor? How much total charge flows? How much total energy is dissipated?

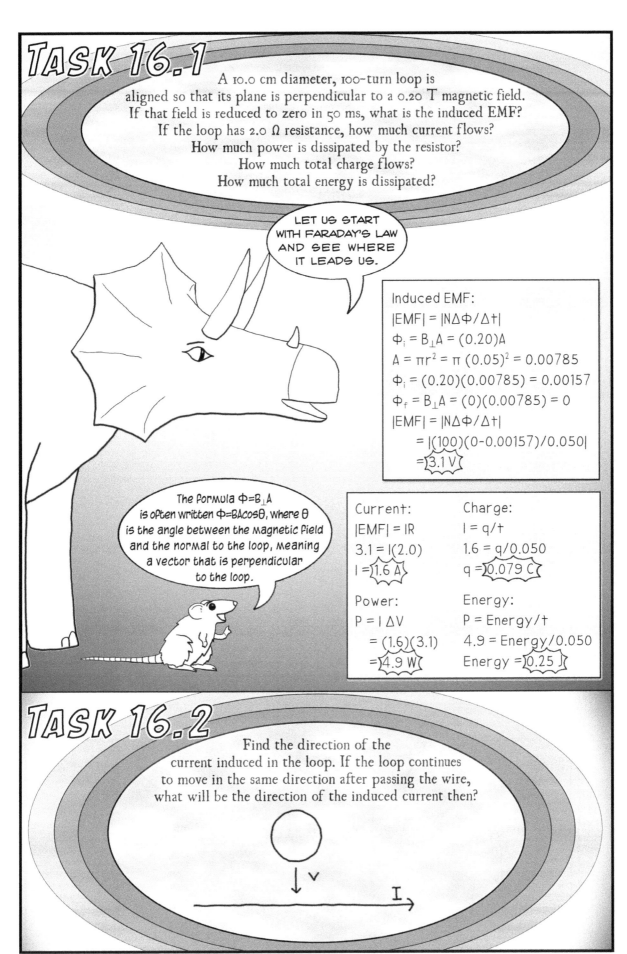

LET US START WITH FARADAY'S LAW AND SEE WHERE IT LEADS US.

Induced EMF:
$|EMF| = |N\Delta\Phi/\Delta t|$
$\Phi_i = B_\perp A = (0.20)A$
$A = \pi r^2 = \pi(0.05)^2 = 0.00785$
$\Phi_i = (0.20)(0.00785) = 0.00157$
$\Phi_f = B_\perp A = (0)(0.00785) = 0$
$|EMF| = |N\Delta\Phi/\Delta t|$
$\quad = |(100)(0-0.00157)/0.050|$
$\quad = 3.1 \text{ V}$

The formula $\Phi = B_\perp A$ is often written $\Phi = BA\cos\theta$, where θ is the angle between the magnetic field and the normal to the loop, meaning a vector that is perpendicular to the loop.

Current:
$|EMF| = IR$
$3.1 = I(2.0)$
$I = 1.6 \text{ A}$

Charge:
$I = q/t$
$1.6 = q/0.050$
$q = 0.079 \text{ C}$

Power:
$P = I\Delta V$
$\quad = (1.6)(3.1)$
$\quad = 4.9 \text{ W}$

Energy:
$P = \text{Energy}/t$
$4.9 = \text{Energy}/0.050$
$\text{Energy} = 0.25 \text{ J}$

TASK 16.2

Find the direction of the current induced in the loop. If the loop continues to move in the same direction after passing the wire, what will be the direction of the induced current then?

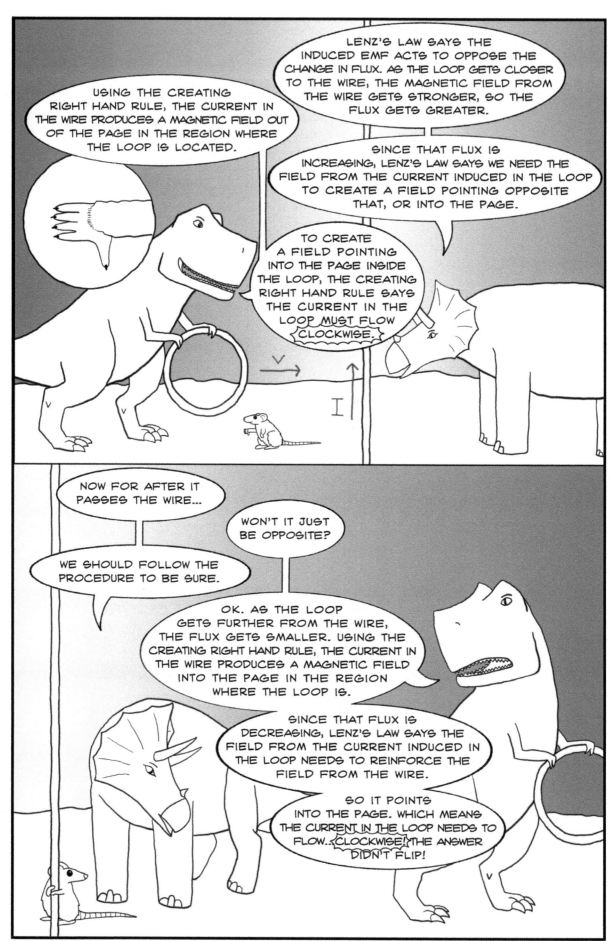

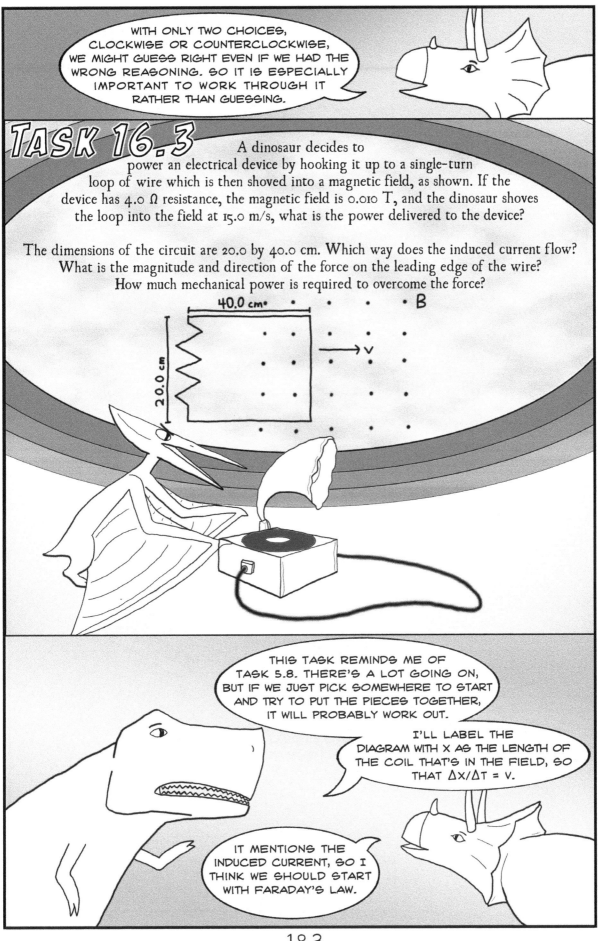

183

IF WE CHANGED THE DIRECTION OF THE MAGNETIC FIELD, THAT WOULD CHANGE THE FLUX THROUGH THE LOOP, RIGHT? CHANGING THE ORIENTATION OF THE COIL SHOULD HAVE THE SAME EFFECT: THE FLUX CHANGES FROM GOING THROUGH THE COIL IN ONE DIRECTION TO GOING THROUGH IN THE OTHER DIRECTION.

I SEE! IF I LABEL ONE SIDE OF THE COIL "FRONT" AND THEN FLIP IT, THEN IF THE FIELD STARTED OUT GOING FROM FRONT TO BACK, FLIPPING THE COIL MAKES IT GO BACK TO FRONT.

B

FRONT

BACK

$$|EMF| = |N\Delta\Phi/\Delta t| = 100|\Delta\Phi/\Delta t|$$
$$\Phi_i = B_\perp A = (0.85)\pi(0.05)^2 = 0.00668$$
Φ_f is the same, but in the opposite direction, so
$$\Phi_f = -0.00668$$
$$|EMF| = |N\Delta\Phi/\Delta t| = 100|[0.00668 - (-0.00668)]/\Delta t|$$
$$= 1.34/\Delta t$$

HOW DO WE KNOW WHICH DIRECTION TO MAKE POSITIVE AND WHICH NEGATIVE?

IT DOES NOT MATTER, SINCE WE TAKE THE ABSOLUTE VALUE.

WE DON'T KNOW ΔT, BUT MAYBE IT WILL CANCEL. LET'S KEEP GOING AND SEE. SINCE WE HAVE RESISTANCE, WE CAN GET CHARGE.

$$|EMF| = 1.34/\Delta t = IR = I(7.0)$$
$$I = 0.191/\Delta t = q/\Delta t$$
Δt's cancel!
$$q = 0.191\ C$$

TASK 16.5

A 25.0 mH inductor is attached in series to a 2.0 Ω resistor as shown. If the battery is connected by opening the switch at t = 0, at what time will the current be 75% of its maximum value?

25 mH

9.0 V

2.0 Ω

186

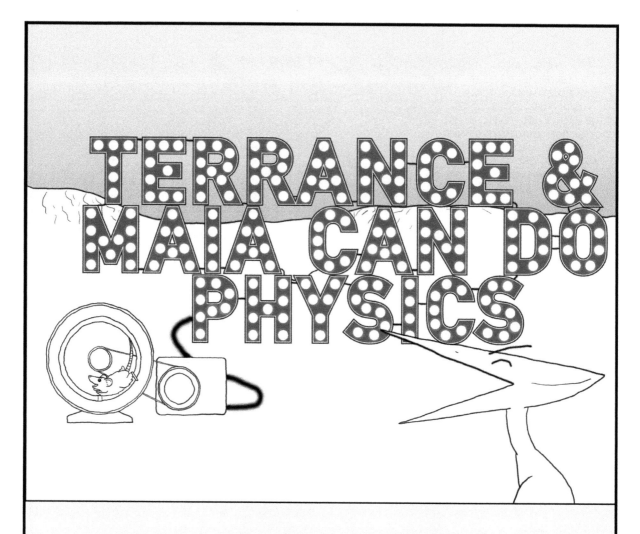

TERRANCE & MAIA CAN DO PHYSICS

EQUATIONS IN THIS CHAPTER

$$\Phi = B_{\perp}A = BA\cos\theta$$

$$|EMF| = |N\Delta\Phi/\Delta t|$$

WORDS OF WISDOM

Lenz's Law: Induced current opposes the change in flux

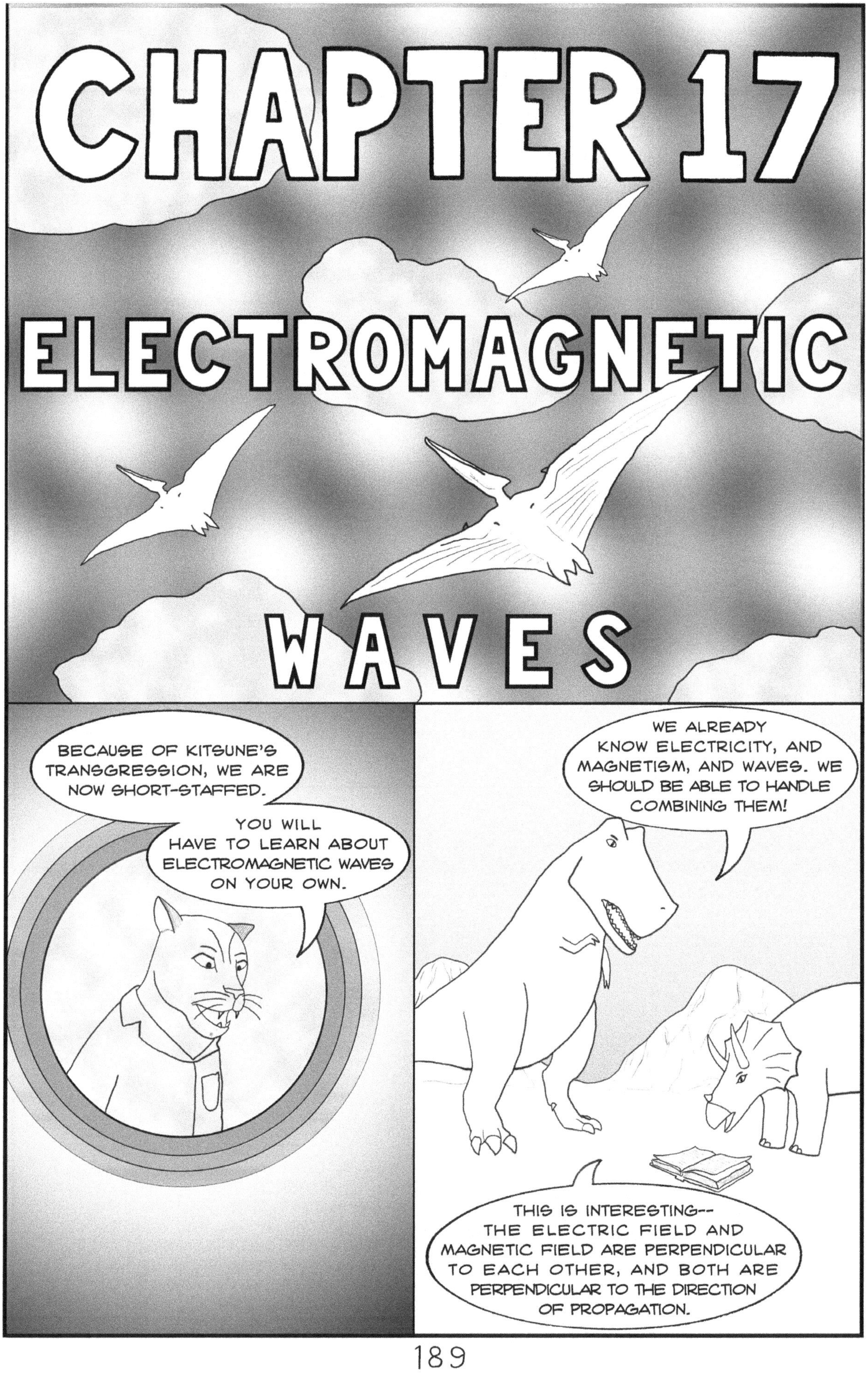
CHAPTER 17
ELECTROMAGNETIC
WAVES
BECAUSE OF KITSUNE'S TRANSGRESSION, WE ARE NOW SHORT-STAFFED.
YOU WILL HAVE TO LEARN ABOUT ELECTROMAGNETIC WAVES ON YOUR OWN.
WE ALREADY KNOW ELECTRICITY, AND MAGNETISM, AND WAVES. WE SHOULD BE ABLE TO HANDLE COMBINING THEM!
THIS IS INTERESTING-- THE ELECTRIC FIELD AND MAGNETIC FIELD ARE PERPENDICULAR TO EACH OTHER, AND BOTH ARE PERPENDICULAR TO THE DIRECTION OF PROPAGATION.

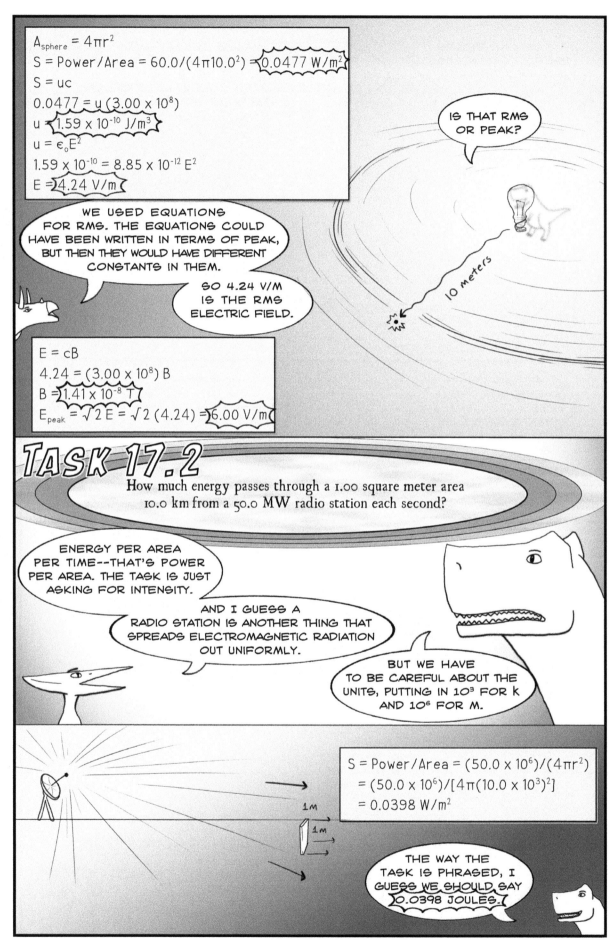

$A_{sphere} = 4\pi r^2$

$S = \text{Power}/\text{Area} = 60.0/(4\pi 10.0^2) = 0.0477 \text{ W/m}^2$

$S = uc$

$0.0477 = u(3.00 \times 10^8)$

$u = 1.59 \times 10^{-10} \text{ J/m}^3$

$u = \epsilon_o E^2$

$1.59 \times 10^{-10} = 8.85 \times 10^{-12} E^2$

$E = 4.24 \text{ V/m}$

IS THAT RMS OR PEAK?

10 meters

WE USED EQUATIONS FOR RMS. THE EQUATIONS COULD HAVE BEEN WRITTEN IN TERMS OF PEAK, BUT THEN THEY WOULD HAVE DIFFERENT CONSTANTS IN THEM.

SO 4.24 V/M IS THE RMS ELECTRIC FIELD.

$E = cB$

$4.24 = (3.00 \times 10^8) B$

$B = 1.41 \times 10^{-8} \text{ T}$

$E_{peak} = \sqrt{2} E = \sqrt{2}(4.24) = 6.00 \text{ V/m}$

TASK 17.2

How much energy passes through a 1.00 square meter area 10.0 km from a 50.0 MW radio station each second?

ENERGY PER AREA PER TIME--THAT'S POWER PER AREA. THE TASK IS JUST ASKING FOR INTENSITY.

AND I GUESS A RADIO STATION IS ANOTHER THING THAT SPREADS ELECTROMAGNETIC RADIATION OUT UNIFORMLY.

BUT WE HAVE TO BE CAREFUL ABOUT THE UNITS, PUTTING IN 10^3 FOR k AND 10^6 FOR M.

1 m

1 m

$S = \text{Power}/\text{Area} = (50.0 \times 10^6)/(4\pi r^2)$

$= (50.0 \times 10^6)/[4\pi(10.0 \times 10^3)^2]$

$= 0.0398 \text{ W/m}^2$

THE WAY THE TASK IS PHRASED, I GUESS WE SHOULD SAY 0.0398 JOULES.

EQUATIONS IN THIS CHAPTER

$$E = cB$$

$$u = \epsilon_o E^2$$

$$S = \text{Power/Area}$$

$$S = uc$$

$$\text{Peak} = \sqrt{2}\,(\text{rms})$$

WORDS OF WISDOM

If a task doesn't state, assume voltage and current are RMS and power is average.

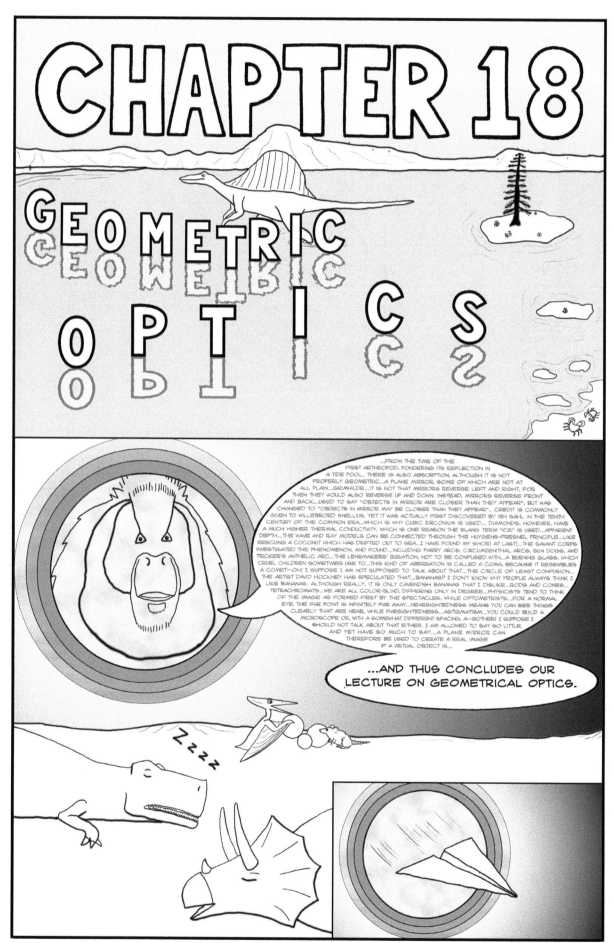

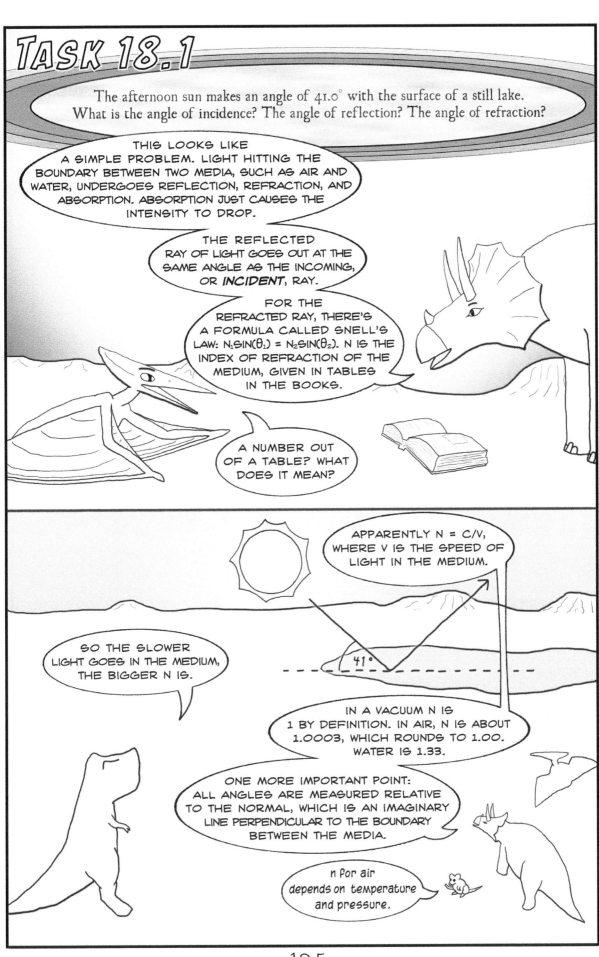

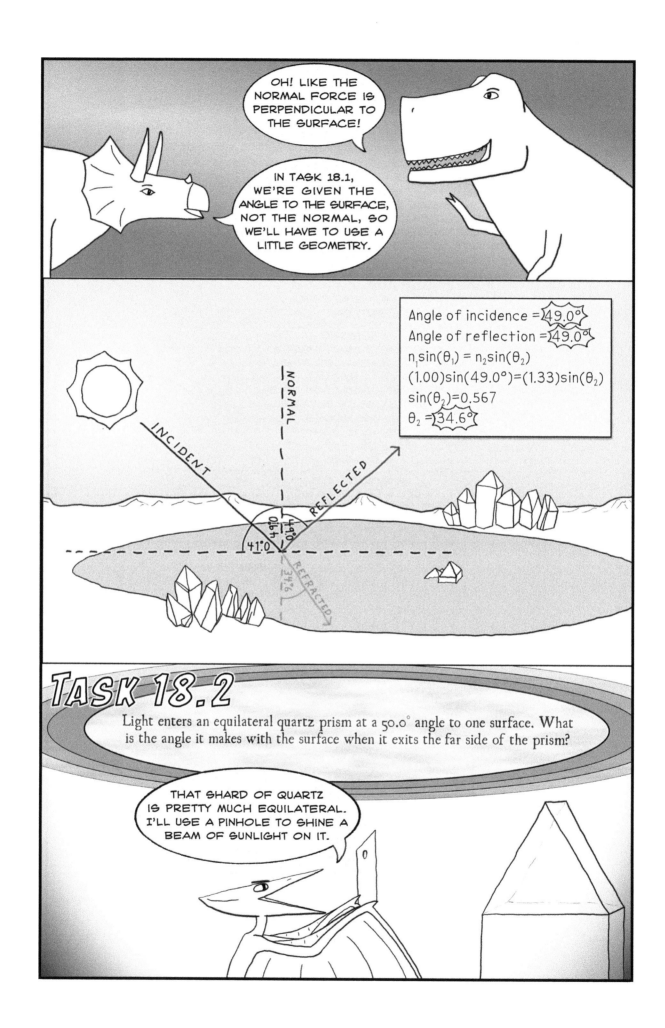

197

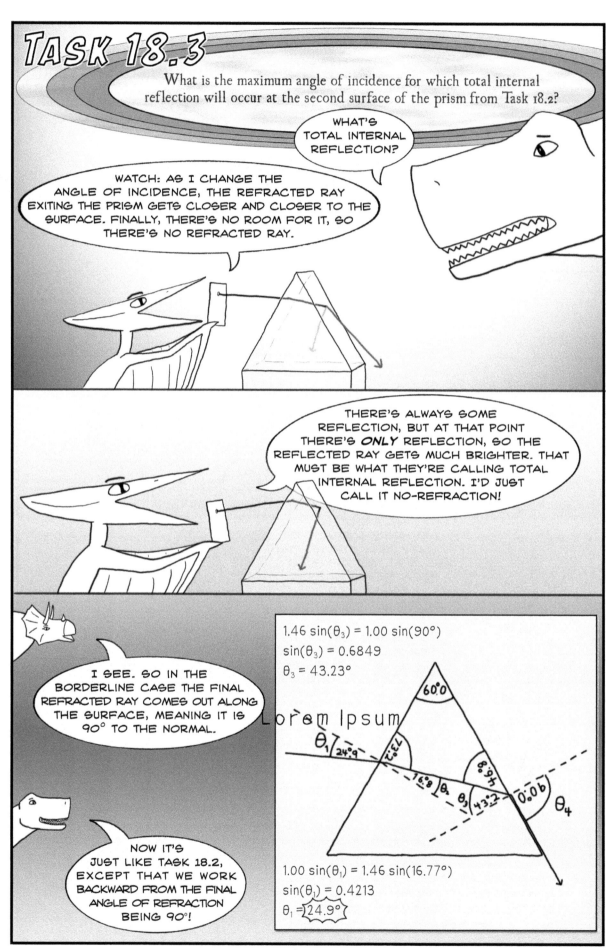

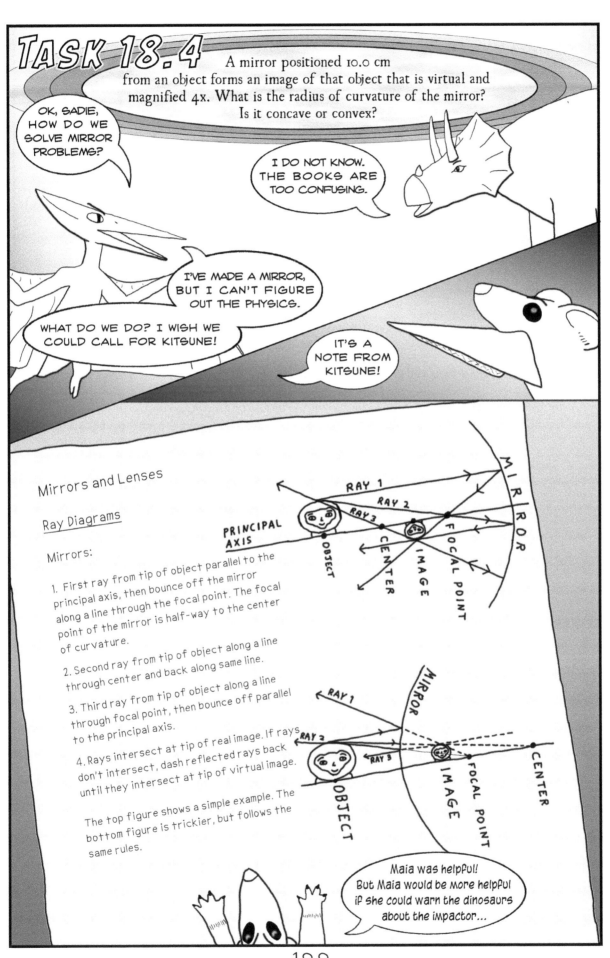

Ray Diagrams (continued)

Converging lens
(thicker in the middle than at top or bottom):

1. First ray from tip of object parallel to the principal axis to central axis of lens, then through the far focal point.

2. Second ray from tip of object straight through center of lens.

3. Third ray from tip of object through near focal point to central axis of lens, then parallel to the principal axis.

4. Rays intersect at tip of real image. If rays don't intersect, dash reflected rays back until they intersect at tip of virtual image.

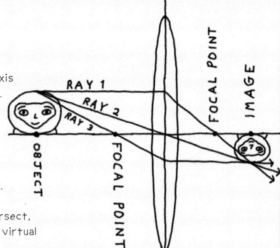

Diverging lens
(thinner in the middle than at top or bottom):

1. First ray from tip of object parallel to the principal axis to central axis of lens, then lined up with the near focal point but continuing forward.

2. Second ray from tip of object straight through center of lens.

3. Third ray from tip of object lined up with far focal point to central axis of lens, then parallel to the principal axis.

4. Dash reflected rays back until they intersect at tip of virtual image.

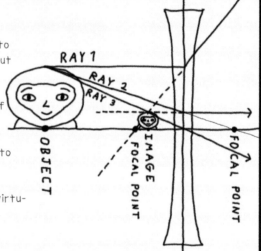

Equations and Sign Conventions

$$1/f = 1/d_i + 1/d_o$$
$$m = h_i/h_o = -d_i/d_o$$

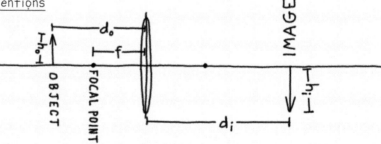

The focal length f is positive for converging lenses and concave mirrors; negative for diverging lenses and convex mirrors

The image distance d_i is positive for a real image, negative for a virtual image

The object distance d_o is always positive for a single lens or mirror

The magnification m is greater than 1 for an enlarged image, less than 1 for a reduced image

The image height h_i is positive for an upright image, negative for an inverted image

The object heigh h_o is always positive for a single lens or mirror

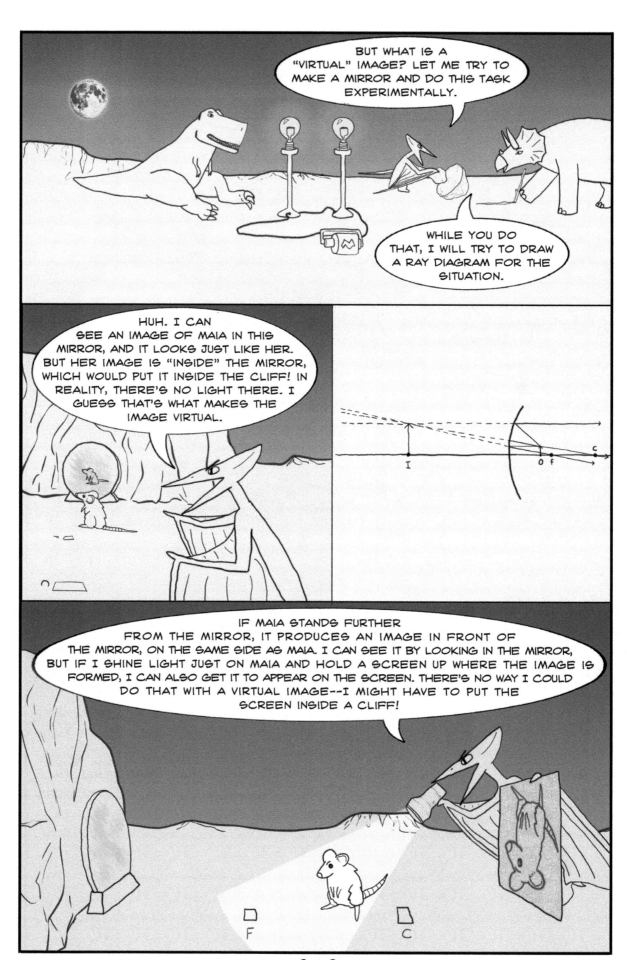

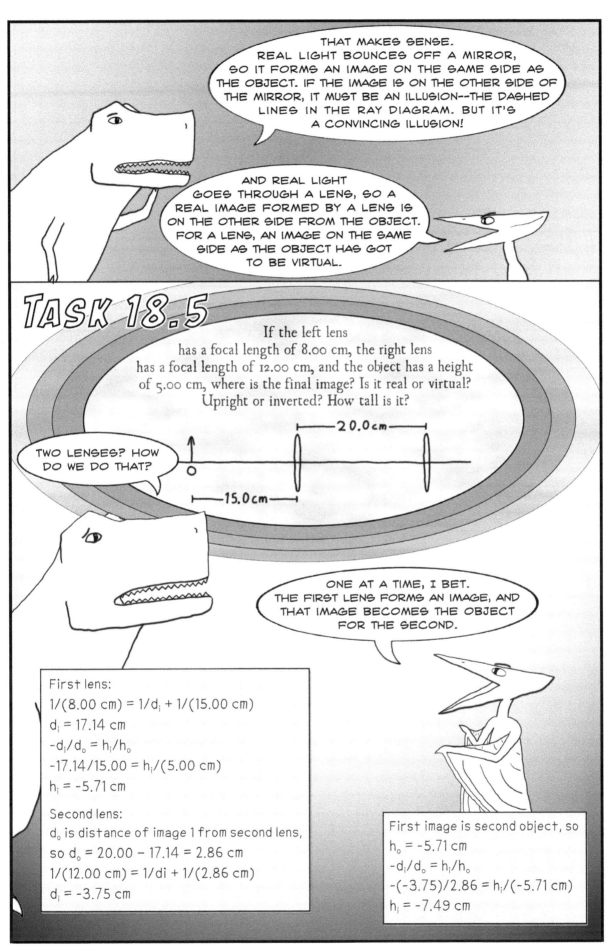

THAT MAKES SENSE. REAL LIGHT BOUNCES OFF A MIRROR, SO IT FORMS AN IMAGE ON THE SAME SIDE AS THE OBJECT. IF THE IMAGE IS ON THE OTHER SIDE OF THE MIRROR, IT MUST BE AN ILLUSION--THE DASHED LINES IN THE RAY DIAGRAM. BUT IT'S A CONVINCING ILLUSION!

AND REAL LIGHT GOES THROUGH A LENS, SO A REAL IMAGE FORMED BY A LENS IS ON THE OTHER SIDE FROM THE OBJECT. FOR A LENS, AN IMAGE ON THE SAME SIDE AS THE OBJECT HAS GOT TO BE VIRTUAL.

TASK 18.5

If the left lens has a focal length of 8.00 cm, the right lens has a focal length of 12.00 cm, and the object has a height of 5.00 cm, where is the final image? Is it real or virtual? Upright or inverted? How tall is it?

TWO LENSES? HOW DO WE DO THAT?

ONE AT A TIME, I BET. THE FIRST LENS FORMS AN IMAGE, AND THAT IMAGE BECOMES THE OBJECT FOR THE SECOND.

First lens:
$1/(8.00 \text{ cm}) = 1/d_i + 1/(15.00 \text{ cm})$
$d_i = 17.14 \text{ cm}$
$-d_i/d_o = h_i/h_o$
$-17.14/15.00 = h_i/(5.00 \text{ cm})$
$h_i = -5.71 \text{ cm}$

Second lens:
d_o is distance of image 1 from second lens,
so $d_o = 20.00 - 17.14 = 2.86 \text{ cm}$
$1/(12.00 \text{ cm}) = 1/di + 1/(2.86 \text{ cm})$
$d_i = -3.75 \text{ cm}$

First image is second object, so
$h_o = -5.71 \text{ cm}$
$-d_i/d_o = h_i/h_o$
$-(-3.75)/2.86 = h_i/(-5.71 \text{ cm})$
$h_i = -7.49 \text{ cm}$

THE FINAL IMAGE IS VIRTUAL, SO IT'S 3.75 CM TO THE LEFT OF THE SECOND LENS. SINCE H_i IS NEGATIVE, IT'S INVERTED, AND 7.49 CM TALL.

TASK 18.6

Repeat Task 18.5, but with the lenses being 10.0 cm apart.

THAT DOES NOT SEEM LIKE AN INTERESTING PROBLEM. IT JUST HAS DIFFERENT NUMBERS. IN FACT, THE FIRST LENS PORTION IS IDENTICAL TO TASK 18.5.

BUT THE IMAGE IS FORMED 17.14 CM TO THE RIGHT OF THE FIRST LENS! HOW IS THAT POSSIBLE? THE SECOND LENS IS IN THE WAY.

ONE OF THESE BOOKS DISCUSSES "VIRTUAL OBJECTS" IN MULTIPLE LENS SYSTEMS. I WOULD GUESS THE IMAGE THAT IS PREVENTED FROM FORMING BECAUSE THE SECOND LENS IS IN THE WAY IS A VIRTUAL OBJECT.

BUT WHAT DO WE DO? KITSUNE'S NOTES DON'T COVER THIS.

PERHAPS THE MATH WILL WORK ANYWAY. LET US CONTINUE WITH THE SOLUTION JUST AS WE DID IN TASK 18.5.

Second lens:
d_o is distance of image 1 from second lens,
so $d_o = 10.00 - 17.14 = -7.14$ cm
$1/(12.00 \text{ cm}) = 1/d_i + 1/(-7.14 \text{ cm})$
$d_i = 4.48$ cm

First image is second object, so
$h_o = -5.71$ cm
$-d_i/d_o = h_i/h_o$
$-(4.48)/(-7.14) = h_i/(-5.71 \text{ cm})$
$h_i = -3.58$ cm

D_i IS POSITIVE, SO THE FINAL IMAGE IS REAL--WEIRD! IT'S 5.71 CM TO THE RIGHT OF THE RIGHT LENS. IT'S INVERTED AND 3.58 CM TALL.

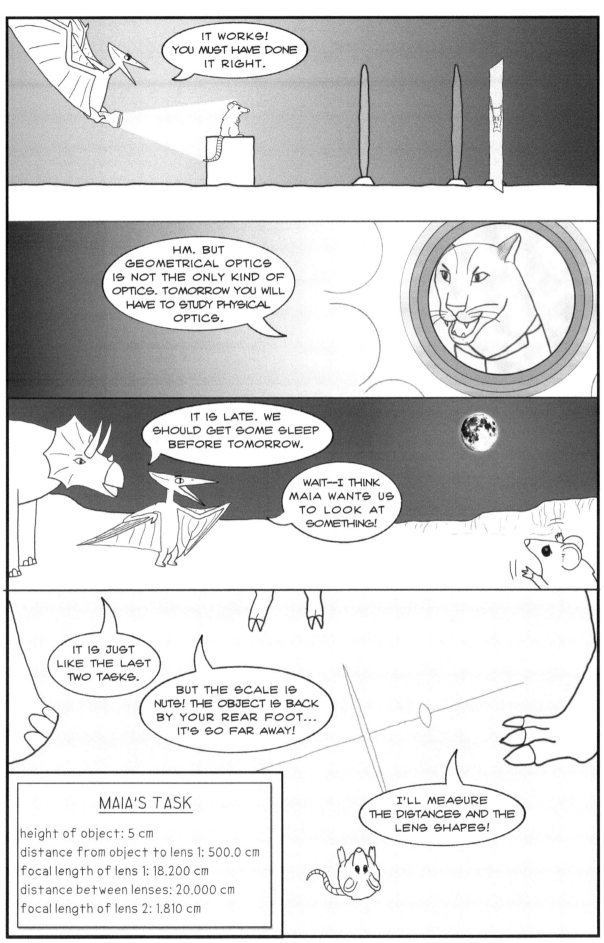

MAIA'S TASK

height of object: 5 cm
distance from object to lens 1: 500.0 cm
focal length of lens 1: 18.200 cm
distance between lenses: 20.000 cm
focal length of lens 2: 1.810 cm

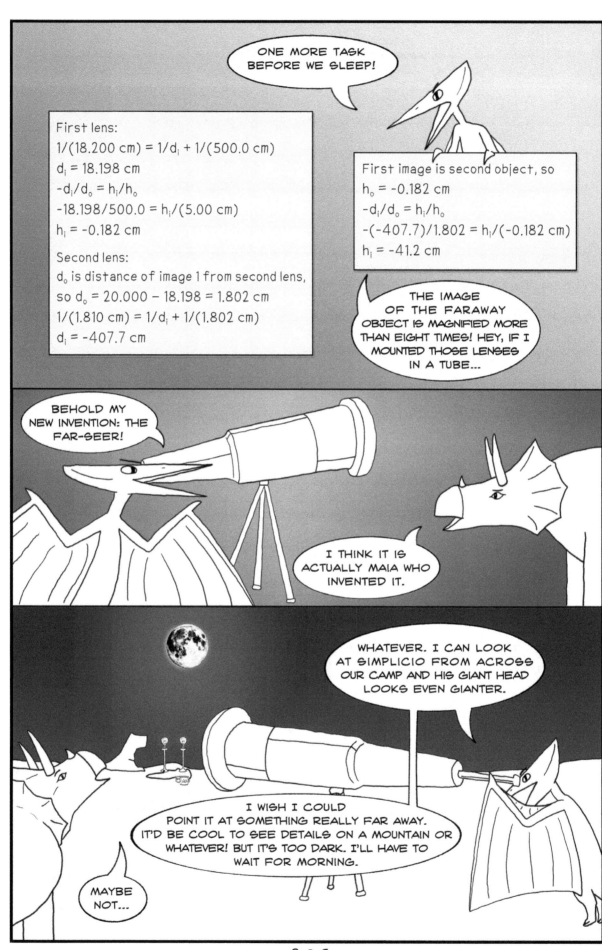

EQUATIONS IN THIS CHAPTER

Reflection: $\theta_r = \theta_i$

Refraction: $n_1 \sin\theta_1 = n_2 \sin\theta_2$; $n = c/v$

$1/f = 1/d_i + 1/d_o$; $m = h_i/h_o = -d_i/d_o$

SIGN CONVENTIONS

The focal length f is positive for converging lenses and concave mirrors

The focal length f is negative for diverging lenses and convex mirrors

The image distance d_i is positive for a real image, negative for a virtual image

The object distance d_o is always positive for a single lens or mirror

The magnification m is greater than 1 for an enlarged image, less than 1 for a reduced image

The image height h_i is positive for an upright image, negative for an inverted image

The object height h_o is always positive for a single lens or mirror

WORDS OF WISDOM

In geometrical optics, all angles are measured
relative to the normal to the surface

Work around the edges of a prism

Real light goes through a lens but bounces off of a mirror

The image of the first lens becomes the object for the second

CHAPTER 19
PHYSICAL OPTICS

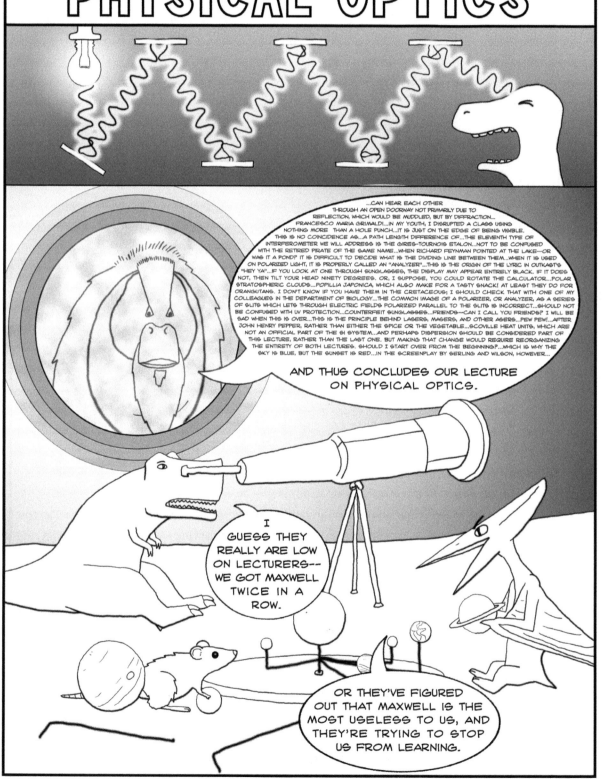

...CAN HEAR EACH OTHER THROUGH AN OPEN DOORWAY NOT PRIMARILY DUE TO REFLECTION, WHICH WOULD BE MUDDLED, BUT BY DIFFRACTION... FRANCESCO MARIA GRIMALDI...IN MY YOUTH, I DISRUPTED A CLASS USING NOTHING MORE THAN A HOLE PUNCH...IT IS JUST ON THE EDGE OF BEING VISIBLE. THIS IS NO COINCIDENCE AS...A PATH LENGTH DIFFERENCE OF...THE ELEVENTH TYPE OF INTERFEROMETER WE WILL ADDRESS IS THE GIRES-TOURNOIS ETALON...NOT TO BE CONFUSED WITH THE RETIRED PIRATE OF THE SAME NAME...WHEN RICHARD FEYNMAN POINTED AT THE LAKE—OR WAS IT A POND? IT IS DIFFICULT TO DECIDE WHAT IS THE DIVIDING LINE BETWEEN THEM...WHEN IT IS USED ON POLARIZED LIGHT, IT IS PROPERLY CALLED AN "ANALYZER"...THIS IS THE ORIGIN OF THE LYRIC IN OUTKAST'S "HEY YA"...IF YOU LOOK AT ONE THROUGH SUNGLASSES, THE DISPLAY MAY APPEAR ENTIRELY BLACK. IF IT DOES NOT, THEN TILT YOUR HEAD NINETY DEGREES. OR, I SUPPOSE, YOU COULD ROTATE THE CALCULATOR...POLAR STRATOSPHERIC CLOUDS...POPILLIA JAPONICA, WHICH ALSO MAKE FOR A TASTY SNACK! AT LEAST THEY DO FOR ORANGUTANS. I DON'T KNOW IF YOU HAVE THEM IN THE CRETACEOUS; I SHOULD CHECK THAT WITH ONE OF MY COLLEAGUES IN THE DEPARTMENT OF BIOLOGY...THE COMMON IMAGE OF A POLARIZER, OR ANALYZER, AS A SERIES OF SLITS WHICH LETS THROUGH ELECTRIC FIELDS POLARIZED PARALLEL TO THE SLITS IS INCORRECT...SHOULD NOT BE CONFUSED WITH UV PROTECTION...COUNTERFEIT SUNGLASSES...FRIENDS—CAN I CALL YOU FRIENDS? I WILL BE SAD WHEN THIS IS OVER...THIS IS THE PRINCIPLE BEHIND LASERS, MASERS, AND OTHER ASERS...PEW PEW!...AFTER JOHN HENRY PEPPER, RATHER THAN EITHER THE SPICE OR THE VEGETABLE...SCOVILLE HEAT UNITS, WHICH ARE NOT AN OFFICIAL PART OF THE SI SYSTEM...AND PERHAPS DISPERSION SHOULD BE CONSIDERED PART OF THIS LECTURE, RATHER THAN THE LAST ONE. BUT MAKING THAT CHANGE WOULD REQUIRE REORGANIZING THE ENTIRETY OF BOTH LECTURES. SHOULD I START OVER FROM THE BEGINNING?...WHICH IS WHY THE SKY IS BLUE, BUT THE SUNSET IS RED...IN THE SCREENPLAY BY SERLING AND WILSON, HOWEVER...

AND THUS CONCLUDES OUR LECTURE ON PHYSICAL OPTICS.

I GUESS THEY REALLY ARE LOW ON LECTURERS-- WE GOT MAXWELL TWICE IN A ROW.

OR THEY'VE FIGURED OUT THAT MAXWELL IS THE MOST USELESS TO US, AND THEY'RE TRYING TO STOP US FROM LEARNING.

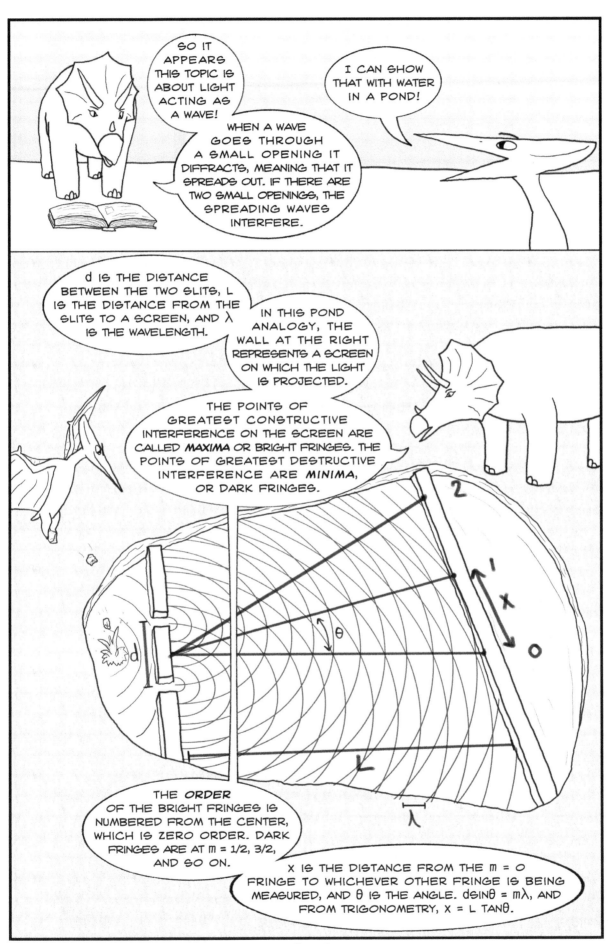

TASK 19.2

a) The second order maximum of a pattern formed when 550 nm light is shone through a diffraction grating with 6000 lines/cm is 3.0 cm from the center of the pattern. How far away is the screen from the grating?

b) 550 nm light is shone through a diffraction grating with 6000 lines/cm. The distance between the first order maximum on each side is 3.0 cm. How far away from the screen is the grating?

c) When 550 nm light is shone through a single slit of width 0.10 mm, the width of the central maximum is 3.0 cm. How far away is the screen from the grating?

Part a:

$1/d$ = 6000 lines/cm

$d = 1/6000$ cm = 1.67×10^{-4} cm = 1.67×10^{-6} m

For diffraction gratings maxima are on the integers, so m = 2.

λ = 550 nm = 550×10^{-9} m

x = 3.0 cm = 3.0×10^{-2} m

$d \sin\theta = m\lambda$

$1.67 \times 10^{-6} \sin\theta = 2(550 \times 10^{-9})$

$\sin\theta = 0.659$; $\theta = 41.2°$

$x = L \tan\theta$

$3.0 \times 10^{-2} = L \tan(41.2°)$

L = 0.034 m

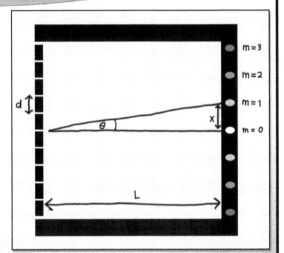

Part b:

$d = 1.67 \times 10^{-6}$ m

$\lambda = 550 \times 10^{-9}$ m

For diffraction gratings maxima are on the integers, so m = 1.

x is distance from center to maximum, which is half the distance from maximum on one side to maximum on the other side.

$x = 1/2 (3.0 \times 10^{-2}$ m) = 1.5×10^{-2} m

$d \sin\theta = m\lambda$

$1.67 \times 10^{-6} \sin\theta = 1(550 \times 10^{-9})$

$\sin\theta = 0.329$; $\theta = 19.2°$

$x = L \tan\theta$

$1.5 \times 10^{-2} = L \tan(19.2°)$

L = 0.043 m

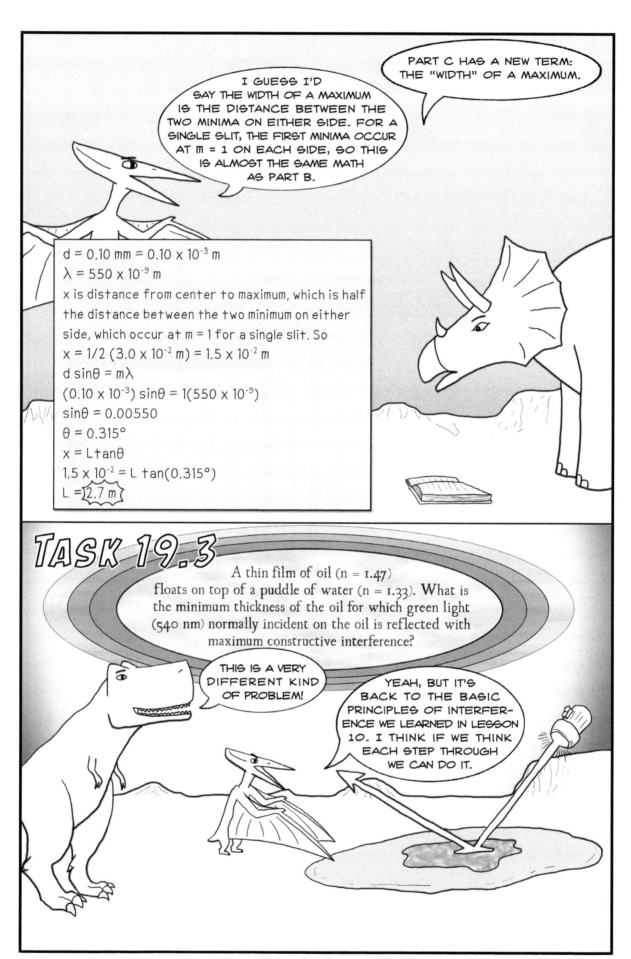

PART C HAS A NEW TERM: THE "WIDTH" OF A MAXIMUM.

I GUESS I'D SAY THE WIDTH OF A MAXIMUM IS THE DISTANCE BETWEEN THE TWO MINIMA ON EITHER SIDE. FOR A SINGLE SLIT, THE FIRST MINIMA OCCUR AT m = 1 ON EACH SIDE, SO THIS IS ALMOST THE SAME MATH AS PART B.

$d = 0.10$ mm $= 0.10 \times 10^{-3}$ m

$\lambda = 550 \times 10^{-9}$ m

x is distance from center to maximum, which is half the distance between the two minimum on either side, which occur at m = 1 for a single slit. So

$x = 1/2 \ (3.0 \times 10^{-2} \text{ m}) = 1.5 \times 10^{-2}$ m

$d \sin\theta = m\lambda$

$(0.10 \times 10^{-3}) \sin\theta = 1(550 \times 10^{-9})$

$\sin\theta = 0.00550$

$\theta = 0.315°$

$x = L\tan\theta$

$1.5 \times 10^{-2} = L \tan(0.315°)$

$L = 2.7$ m

TASK 19.3

A thin film of oil (n = 1.47) floats on top of a puddle of water (n = 1.33). What is the minimum thickness of the oil for which green light (540 nm) normally incident on the oil is reflected with maximum constructive interference?

THIS IS A VERY DIFFERENT KIND OF PROBLEM!

YEAH, BUT IT'S BACK TO THE BASIC PRINCIPLES OF INTERFERENCE WE LEARNED IN LESSON 10. I THINK IF WE THINK EACH STEP THROUGH WE CAN DO IT.

TASK 19.4

What fraction of the original intensity gets through when:

a) unpolarized light is passed through two polarizers which are at 90° to each other?

b) a third polarizer is placed between the two from part a, at an angle of 30° to the first one?

c) a third polarizer is placed after the two from part a, at an angle of 30° to the first one?

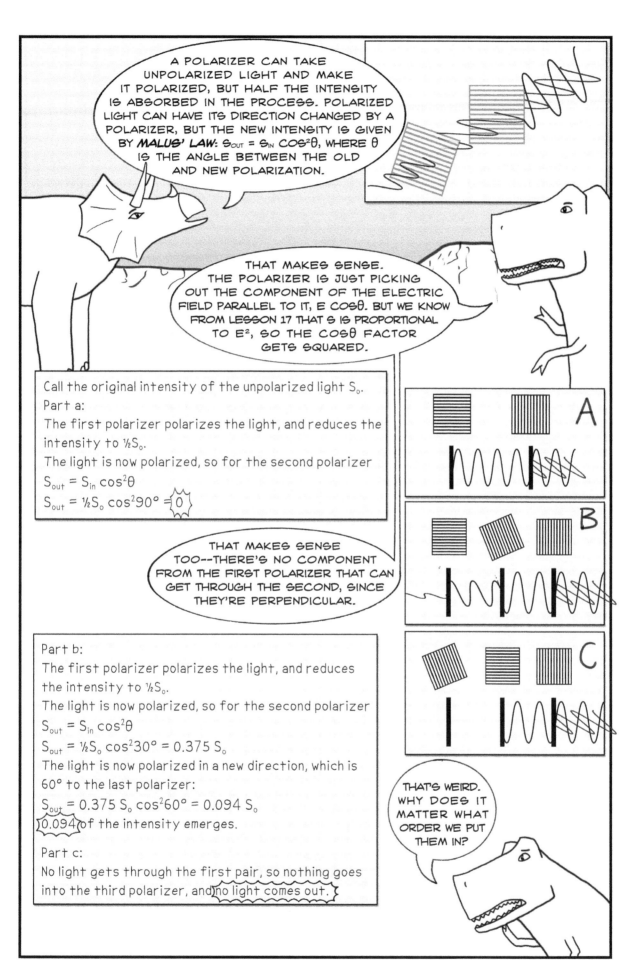

A POLARIZER CAN TAKE UNPOLARIZED LIGHT AND MAKE IT POLARIZED, BUT HALF THE INTENSITY IS ABSORBED IN THE PROCESS. POLARIZED LIGHT CAN HAVE ITS DIRECTION CHANGED BY A POLARIZER, BUT THE NEW INTENSITY IS GIVEN BY *MALUS' LAW*: $S_{out} = S_{in} \cos^2\theta$, WHERE θ IS THE ANGLE BETWEEN THE OLD AND NEW POLARIZATION.

THAT MAKES SENSE. THE POLARIZER IS JUST PICKING OUT THE COMPONENT OF THE ELECTRIC FIELD PARALLEL TO IT, $E \cos\theta$. BUT WE KNOW FROM LESSON 17 THAT S IS PROPORTIONAL TO E^2, SO THE $\cos\theta$ FACTOR GETS SQUARED.

Call the original intensity of the unpolarized light S_o.

Part a:

The first polarizer polarizes the light, and reduces the intensity to $\frac{1}{2}S_o$.

The light is now polarized, so for the second polarizer
$S_{out} = S_{in} \cos^2\theta$
$S_{out} = \frac{1}{2}S_o \cos^2 90° = 0$

THAT MAKES SENSE TOO—THERE'S NO COMPONENT FROM THE FIRST POLARIZER THAT CAN GET THROUGH THE SECOND, SINCE THEY'RE PERPENDICULAR.

Part b:

The first polarizer polarizes the light, and reduces the intensity to $\frac{1}{2}S_o$.

The light is now polarized, so for the second polarizer
$S_{out} = S_{in} \cos^2\theta$
$S_{out} = \frac{1}{2}S_o \cos^2 30° = 0.375\ S_o$

The light is now polarized in a new direction, which is 60° to the last polarizer:
$S_{out} = 0.375\ S_o \cos^2 60° = 0.094\ S_o$
0.094 of the intensity emerges.

Part c:

No light gets through the first pair, so nothing goes into the third polarizer, and no light comes out.

THAT'S WEIRD. WHY DOES IT MATTER WHAT ORDER WE PUT THEM IN?

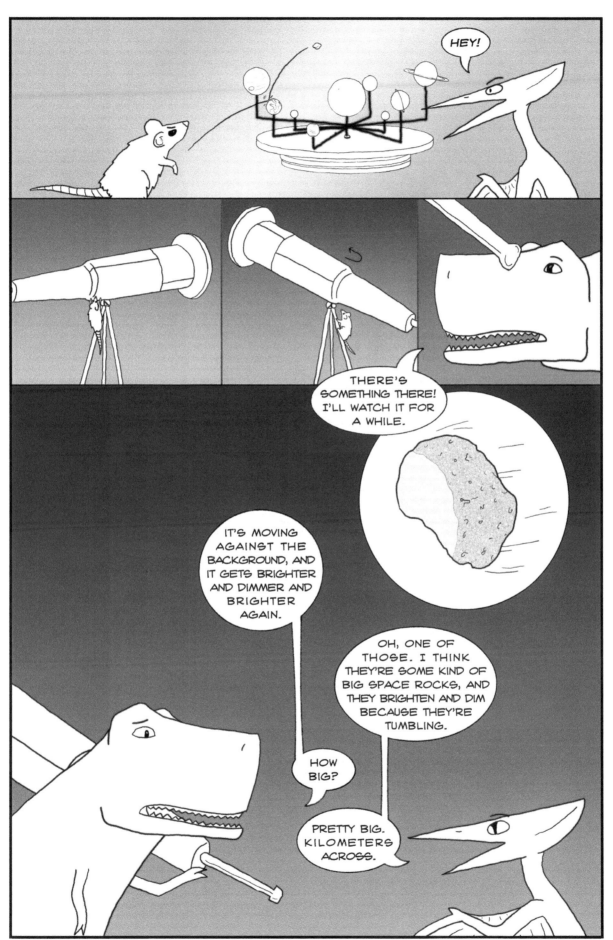

EQUATIONS IN THIS CHAPTER

$d\sin(\theta) = m\lambda$; $x = L\tan(\theta)$

Double slit/diffraction grating: $m = 0, 1, 2, 3, ...$ gives maxima; d is spacing between slits

Single slit: $m = 0, 1/2, 3/2, 5/2, ...$ gives maxima; d is width of slit

Unpolarized Incident Light: $S_{out} = \frac{1}{2} S_{in}$; Malus's Law: $S_{out} = S_{in}\cos^2(\theta)$

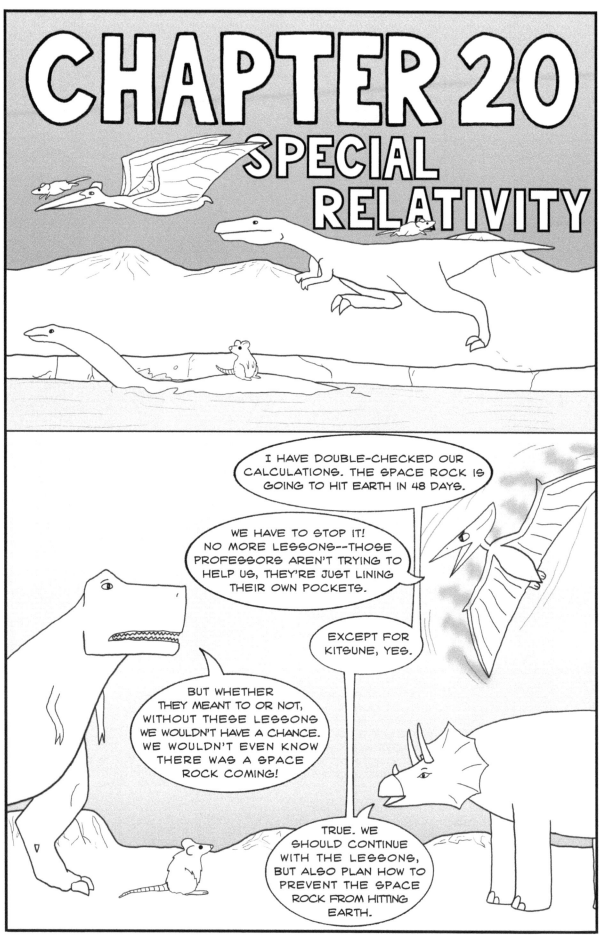

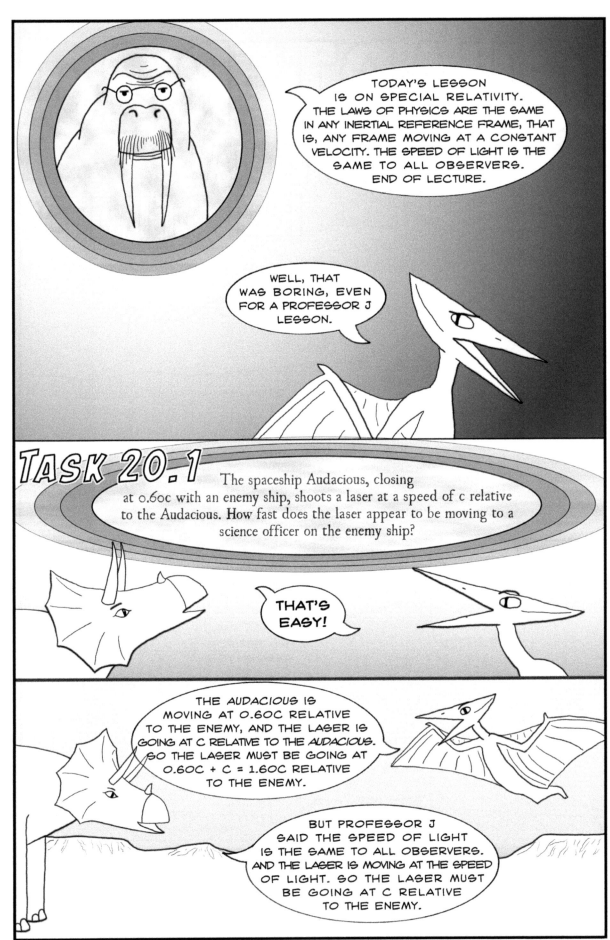

BOTH ARGUMENTS MAKE SENSE. BUT SADIE'S MATCHES WHAT PROFESSOR J SAID, SO SOMETHING MUST BE WRONG WITH TERRANCE'S. COULD YOU GIVE US YOUR SOLUTION STEP BY STEP, TERRANCE?

SURE. TAKE THREE OBJECTS, A, B, AND C. CALL THE VELOCITY OF A RELATIVE TO B V_{AB}, AND B RELATIVE TO C V_{BC}. THE VELOCITY OF A RELATIVE TO C IS THEN $V_{AC} = V_{AB} + V_{BC}$.

SO IF I'M DIVING TOWARD THE GROUND AT 8 M/S, AND I THROW A ROCK DOWN AT 5 M/S RELATIVE TO ME, IT STARTS OUT AT 13 M/S RELATIVE TO THE GROUND.

8 m/s

5 m/s

13 m/s

BY THE WAY: $V_{BA} = -V_{AB}$. THE ORDER OF THE LETTERS MATTERS.

The books call this "Galilean relativity." But since they are from the future, Terrance came up with it first! So Maia thinks it's OK to call it "Terrance's relativity."

TERRANCE'S RELATIVITY FORMULA WORKS FINE FOR THE SPEEDS WE GO AT. BUT HOW DO WE KNOW IT WORKS FOR VERY FAST THINGS?

THE BOOKS GIVE THAT FORMULA, BUT THEY ALSO GIVE ANOTHER. IT HAS A LOT OF CONFUSING SYMBOLS.

PART OF IT LOOKS LIKE TERRANCE'S EQUATION! USING HIS SYMBOLS, THE BOOK EQUATION IS $V_{AC} = (V_{AB} + V_{BC})/(1+V_{AB}V_{BC}/C^2)$.

AS LONG AS V_{AB} OR V_{BC} IS MUCH LESS THAN C, IT GIVES NEARLY THE SAME ANSWER AS TERRANCE'S EQUATION.

SO TERRANCE'S EQUATION WORKS AT LOW SPEEDS, BUT MUST BE MODIFIED AT HIGH SPEEDS.

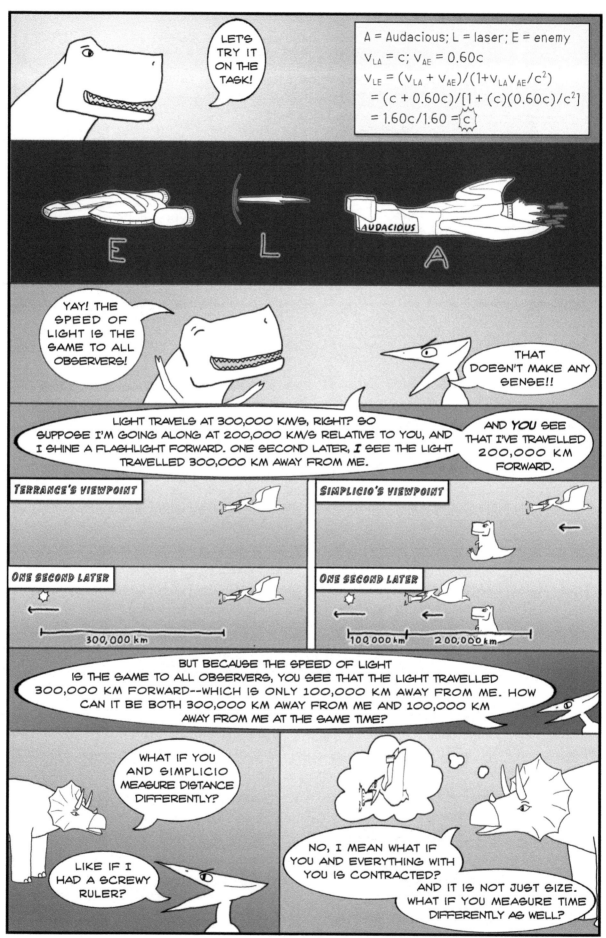

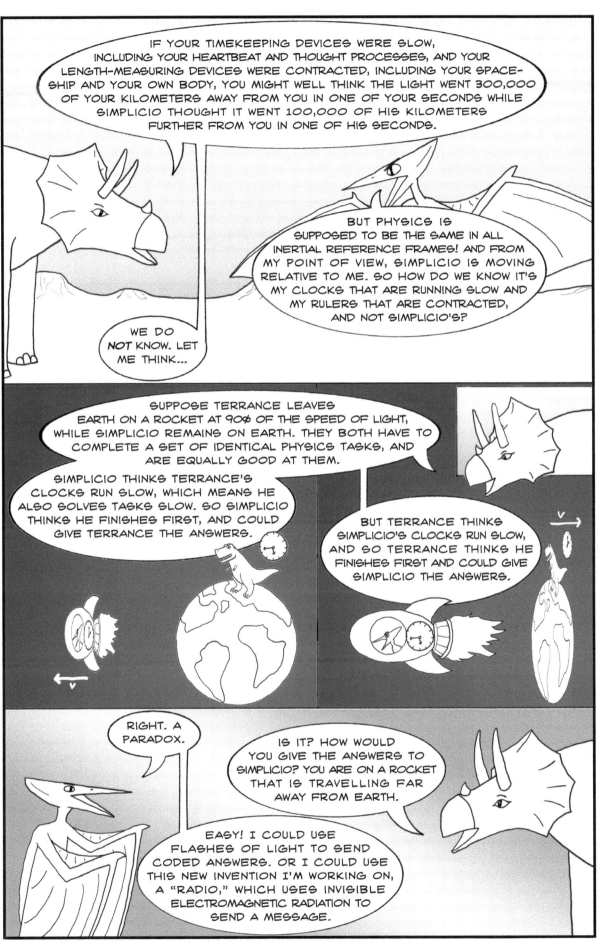

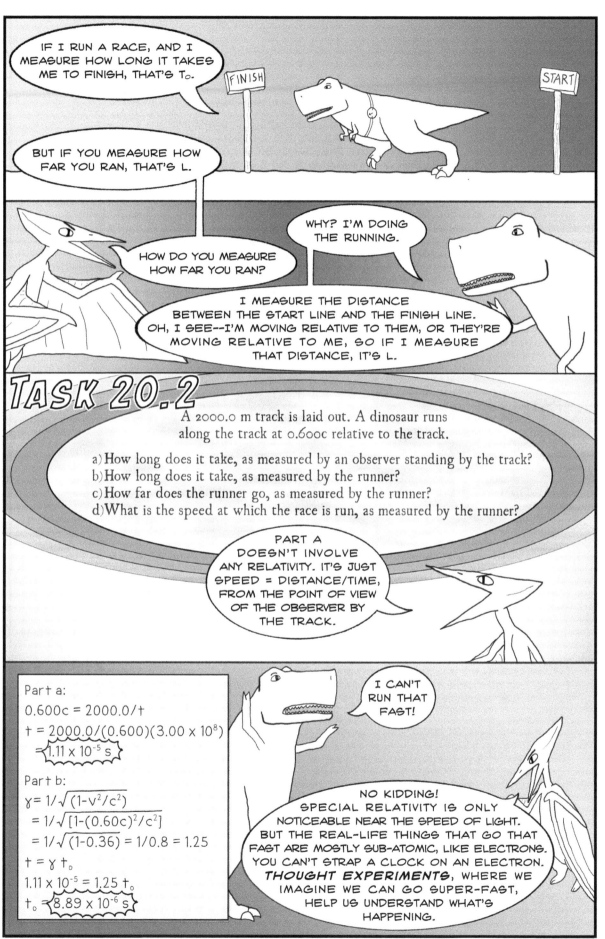

Part c:
$L_o = 2000.0$
$L = L_o/\gamma = 2000.0/1.25$
$= 1600.0 \text{ m}$

Part d:
Speed = distance/time
$= 1600.0/(8.89 \times 10^{-6})$
$= 1.80 \times 10^8 \text{ m/s}$

I JUST NOTICED SOMETHING! 1.80×10^8 M/S = 0.60C, WHICH IS ALSO WHAT THE TRACKSIDE OBSERVER THOUGHT. THEY DISAGREE ON THE DISTANCE AND THE TIME. BUT THEY AGREE ON THE RELATIVE SPEED OF THE RUNNER AND THE TRACKSIDE OBSERVER. THAT SEEMS RIGHT.

TASK 20.3

A spacecraft travels directly from Earth to Alpha Centauri, which is 4.30 light years distant as measured by scientists on Earth. The astronauts on the spacecraft age 5.00 years in the process.

a) What is the speed of the spacecraft relative to the Earth?
b) How long does it take for the spacecraft to get from Earth to Alpha Centauri, as measured by scientists on Earth?

WHAT'S A LIGHT YEAR?

IT MUST BE THE DISTANCE LIGHT TRAVELS IN A YEAR.

SO C = 1 LIGHT YEAR/YEAR. IF WE USE THAT, I DO NOT THINK WE NEED TO CONVERT TO SI UNITS FOR THIS TASK.

5.00 YEARS MUST BE THE TIME MEASURED BY THE ASTRONAUTS, T_o. 4.30 LIGHT YEARS IS MEASURED BY THE SCIENTISTS, L_o. THE SPEED IS THE SAME WHETHER THE SCIENTISTS OR ASTRONAUTS CALCULATE IT. BUT WE DON'T HAVE BOTH SPEED AND DISTANCE FOR EITHER OF THEM.

PERHAPS WE SHOULD TRY IT FROM ONE OF THE POINTS OF VIEW AND SEE IF THE MATHEMATICS WORKS OUT.

From the scientists' viewpoint:
Speed = distance/time
Distance = L_o = 4.30 l.y.
Time = $t = \gamma t_o = \gamma (5.00 \text{ y})$
$v = 4.30 \text{ l.y.}/(\gamma 5.00 \text{ y})$
$v\gamma = 0.860 \text{ l.y./y} = 0.860c$
$v[1/\sqrt{(1-v^2/c^2)}] = 0.860c$
$v = (0.860c)\sqrt{(1-v^2/c^2)}$
$v^2 = 0.740 c^2 (1 - v^2/c^2)$
$\quad = 0.740 c^2 - 0.740 v^2$
$1.740 v^2 = 0.740 c^2$
$v^2 = 0.426 c^2$
$v = 0.652 c$

LET'S TRY IT FROM THE ASTRONAUTS' VIEWPOINT TOO, JUST TO MAKE SURE IT WORKS OUT THE SAME.

From the astronauts' viewpoint:

Speed = distance/time

Distance = $L = L_o/\gamma = 4.30$ l.y./γ

Time = $t_o = 5.00$ y

$v = (4.30$ l.y./$\gamma) / 5.00$ y

$= 4.30$ l.y./$(\gamma\ 5.00$ y)

This is the same equation derived for the scientists, so it has the same solution.

NOW PART B IS EASY! WE JUST SOLVE FOR T.

$t = \gamma t_o$

$\gamma = 1/\sqrt{(1-v^2/c^2)}$

$= 1/\sqrt{[1-(0.652c)^2/c^2]} = 1.319$

$t = (1.319)(5.00$ y$) =$ 6.59 years

TASK 20.4

The Audacious, fleeing from an enemy warship at 0.70c relative to the warship, launches a missile, which approaches the warship at 0.50c. What is the speed of the missile relative to the Audacious?

WE SHOULD SKETCH THIS. LET US DRAW THE **AUDACIOUS** GOING LEFT AND THE MISSILE TO THE RIGHT. I WILL LABEL RIGHT POSITIVE.

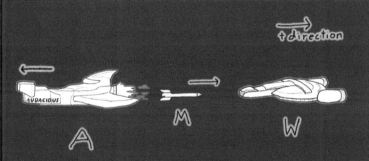

+ direction

$v_{AW} = -0.70c$

$v_{MW} = 0.50c$

$v_{MA} = (v_{MW} + v_{WA})/(1+v_{MW}v_{WA}/c^2)$

$v_{WA} = -v_{AW} = -(-0.70c) = 0.70c$

$v_{MA} = (0.70c+0.50c)/$

$[1+(0.70c)(0.50c)/c^2]$

$= 1.20c/1.35 =$ 0.89c

THERE'S STILL SOME-THING I DON'T GET.

THE RELATIVISTIC ADDITION OF VELOCITY FORMULA WORKS OUT SO THAT NOTHING IS EVER GOING FASTER THAN THE SPEED OF LIGHT. BUT WHAT IF A BIRD PASSES ME AT 0.99C, AND I GIVE IT A LITTLE SHOVE. WHY CAN'T I ACCELERATE IT PAST LIGHT SPEED?

HMM. F = MA, SO...

AT LOW SPEEDS, E = GAMMA MC²
= (1 + 1/2 V²/C²) MC² = MC² + 1/2 MV². THAT KIND OF LOOKS LIKE KE = 1/2 MV²,
BUT WITH AN EXTRA MC² ADDED IN.

AT LOW SPEEDS, WE KNOW THAT KINETIC ENERGY IS ½ MV², SO MC² MUST BE A KIND OF POTENTIAL ENERGY ASSOCIATED WITH THE MASS.

AT LOW SPEEDS, MC² WOULD BE GINORMOUS COMPARED TO 1/2 MV²! IF THERE WAS A WAY TO TURN THAT "REST ENERGY" INTO OTHER KINDS OF ENERGY, THAT WOULD BE AWESOME!

THE BOOK SAYS THAT E = GAMMA MC² INCLUDES ALL FORMS OF ENERGY. BUT WHAT ABOUT THE KINDS OF POTENTIAL ENERGY WE KNOW ABOUT, SUCH AS GRAVITATIONAL OR ELECTRICAL ENERGY?

I BET IF AN OBJECT HAS THOSE ITS MASS WOULD CHANGE A LITTLE BIT.

IT APPEARS SO-- BUT THE MC² IT STARTS WITH WOULD BE SO LARGE THAT THE CHANGE IN MASS WOULD BE VERY HARD TO DETECT IN ORDINARY CIRCUMSTANCES.

TASK 20.5

For each of the following, indicate whether it a) must refer to a kinetic energy, or could refer to kinetic or total energy. b) If they could refer to either, for which is there a difference to two significant figures? c) For which cases is gamma greater than 1.01? The mass of an electron is 9.11 x 10⁻³¹ kg, and a proton is 1.67 x 10⁻²⁷ kg. 200 eV electron, 200 eV proton, 200 keV electron, 200 keV proton, 2 MeV electron, 2 MeV proton, 2000 MeV electron, 2000 MeV proton

IF THE ENERGY GIVEN IS LESS THAN MC², IT MUST BE A KINETIC ENERGY, OTHERWISE IT COULD BE EITHER. SO WE NEED TO CALCULATE MC² FOR AN ELECTRON AND A PROTON.

Part a:
Rest energy of an electron:
$E = mc^2 = (9.11 \times 10^{-31})(3.00 \times 10^8)^2 = $ J $= 5.12 \times 10^5$ eV $= 512$ keV
Rest energy of a proton:
$E = mc^2 = (1.67 \times 10^{-27})(3.00 \times 10^8)^2 = $ J $= 9.39 \times 10^8$ eV $= 939$ MeV
Since these are less than the rest energy, they must refer to a kinetic energy: 200 eV electron, 200 keV electron, 200 eV proton, 200 keV proton, 2 MeV proton

FOR PART B, THE RESTING ENERGY OF THE ELECTRON IS 512 KEV = 0.512 MEV, WHICH WOULD NOT MAKE A DIFFERENCE IN THE SECOND SIGNIFICANT FIGURE OF 2000 MEV. THAT IS THE ONLY ONE FOR WHICH THERE IS A NEGLIGIBLE DIFFERENCE BETWEEN TOTAL ENERGY AND KINETIC ENERGY.

TOTAL ENERGY IS GAMMA MC^2. SO IF GAMMA IS GREATER THAN 1, THE TOTAL ENERGY IS LARGER THAN THE REST ENERGY, WHICH IS ANOTHER WAY OF SAYING THE KINETIC ENERGY IS SIGNIFICANT COMPARED TO THE REST ENERGY.

Part c:
200 eV kinetic energy electron:
$E_{total} = 5.12 \times 10^5 + 200 = 5.12 \times 10^5$ eV
$\gamma = 5.12 \times 10^5 / 5.12 \times 10^5 = 1.00$.

200 keV kinetic energy electron:
$E_{total} = 5.12 \times 10^5 + 2.00 \times 10^5 = 7.12 \times 10^5$ eV
$\gamma = 7.12 \times 10^5 / 5.12 \times 10^5 = 1.39$,
which is greater than 1.01.

All the higher energy electrons would also have γ larger than 1.

FOR THE PROTON, EVEN 2 MEV ISN'T QUITE ENOUGH, SINCE GAMMA IS (939+2)/939 = 1.002. ONLY THE 2000 MEV PROTON HAS GAMMA GREATER THAN 1.01.

I'M CURIOUS. THE TASK DIDN'T ASK FOR IT, BUT WHAT SPEED CORRESPONDS TO GAMMA = 1.01?

$\gamma = 1/sqrt(1-v^2/c^2)$
$1.01 = 1/sqrt(1-v^2/c^2)$
$1.01^2 = 1.02 = 1/(1-v^2/c^2)$
$0.98 = (1-v^2/c^2))$
$v^2/c^2 = 0.02$
$v/c = 0.14$
$v = 0.14c = 0.14(3.0 \times 10^8 \, m/s)$
$= 4.2 \times 10^7 \, m/s$

SO RELATIVITY DOESN'T START BECOMING NOTICEABLE UNTIL SPEEDS ARE FAR MORE THAN 10^7 M/S. EVEN VERY FAST EVERY DAY OBJECTS DON'T SHOW NOTICEABLE RELATIVISTIC EFFECTS.

EQUATIONS IN THIS CHAPTER

$$v_{AC} = (v_{AB} + v_{BC})/(1 + v_{AB}v_{BC}/c^2)$$

$$v_{BA} = -v_{AB}$$

$$\gamma = 1/\sqrt{1 - v^2/c^2}$$

$$t = \gamma t_o$$

$$L = L_o/\gamma$$

$$F = \gamma ma$$

$$E = \gamma mc^2$$

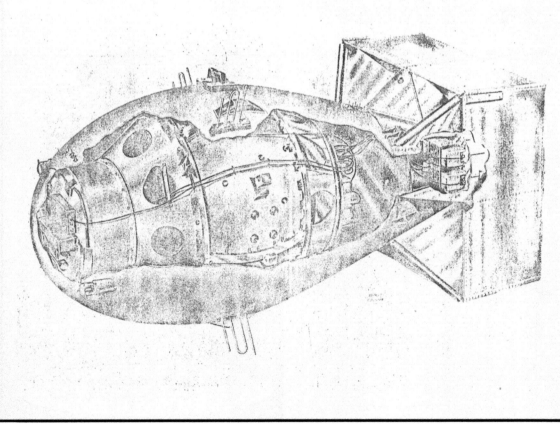

CHAPTER 21
QUANTUM PHYSICS
OH GOOD! IT'S PROFESSOR J. BESIDES KITSUNE, I LIKE HIS LECTURES BEST.
I AM HERE TO DELIVER YOUR FINAL LECTURE, ON MODERN PHYSICS.
I DO NOT LIKE THIS LECTURE, NOT BECAUSE OF THE TOPIC, BUT BECAUSE IT IS SEVERAL LITTLE TOPICS, EACH OF WHICH MUST BE EXPLAINED, REQUIRING MANY WORDS IN TOTAL. THEREFORE I WILL DELIVER IT ONE TOPIC AT A TIME, RETURNING WHEN IT IS TIME FOR THE NEXT TOPIC. END OF RANT.
TOPIC 1: THE SMALLEST PACKAGE, OR QUANTUM, OF ELECTROMAGNETIC RADIATION IS THE PHOTON. THE ENERGY OF A SINGLE PHOTON IS HF, WHERE H IS PLANCK'S CONSTANT, 6.63×10^{-34}JS, AND F IS FREQUENCY. END OF FIRST TOPIC.

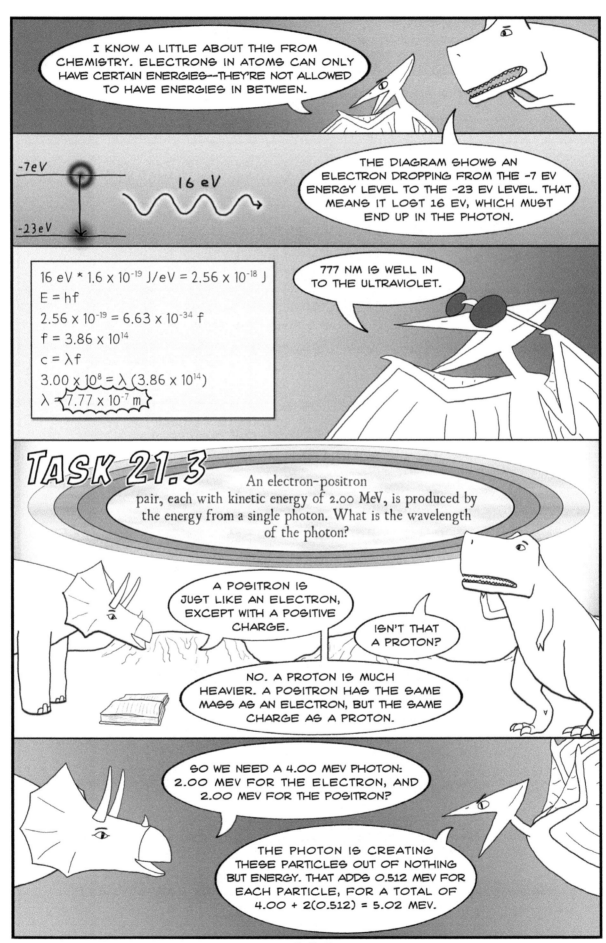

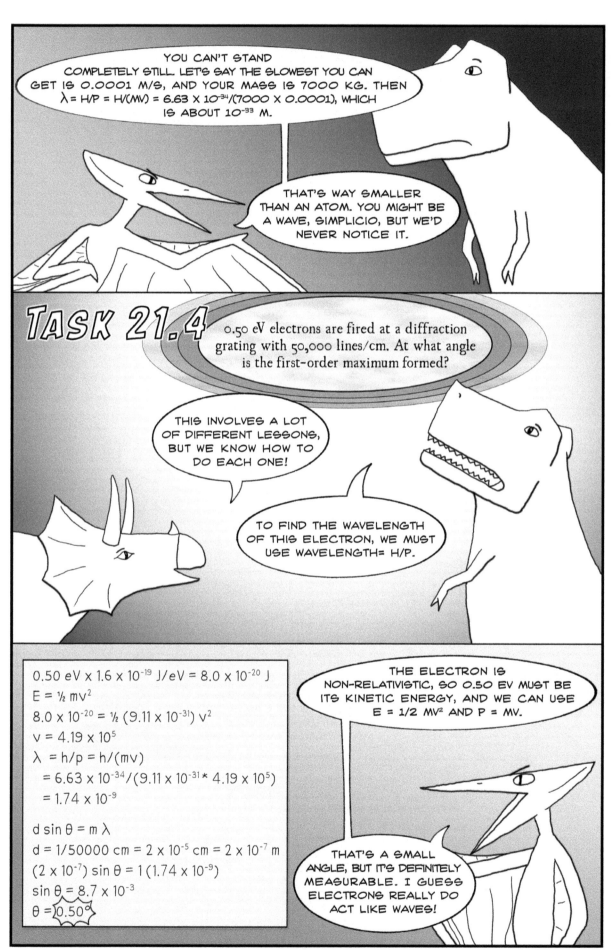

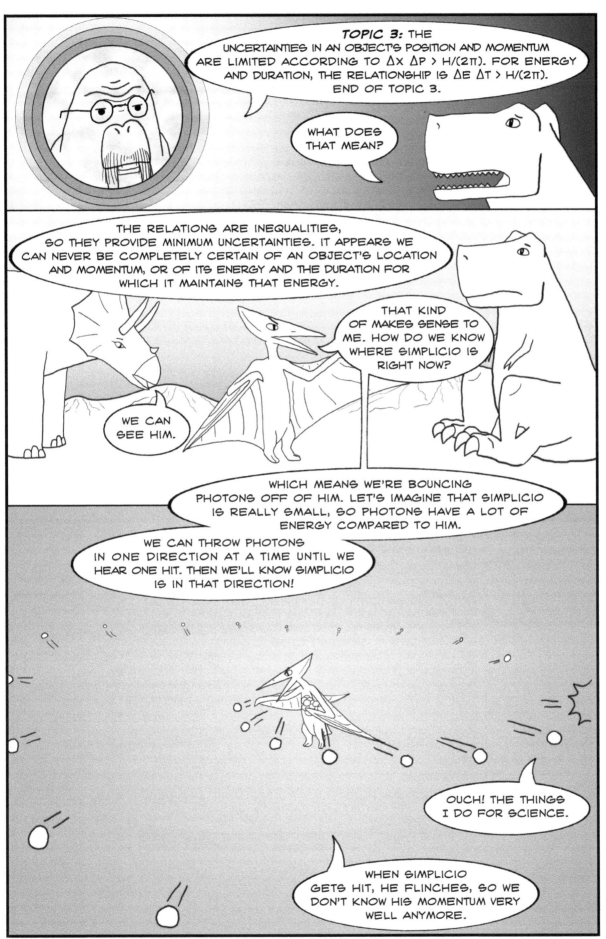

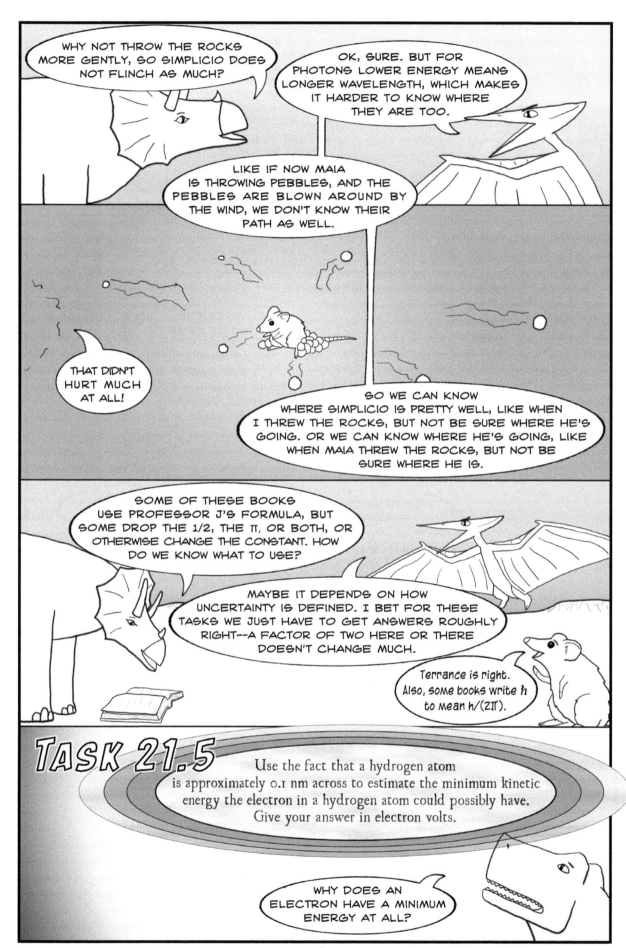

WHY NOT THROW THE ROCKS MORE GENTLY, SO SIMPLICIO DOES NOT FLINCH AS MUCH?

OK, SURE. BUT FOR PHOTONS LOWER ENERGY MEANS LONGER WAVELENGTH, WHICH MAKES IT HARDER TO KNOW WHERE THEY ARE TOO.

LIKE IF NOW MAIA IS THROWING PEBBLES, AND THE PEBBLES ARE BLOWN AROUND BY THE WIND, WE DON'T KNOW THEIR PATH AS WELL.

THAT DIDN'T HURT MUCH AT ALL!

SO WE CAN KNOW WHERE SIMPLICIO IS PRETTY WELL, LIKE WHEN I THREW THE ROCKS, BUT NOT BE SURE WHERE HE'S GOING. OR WE CAN KNOW WHERE HE'S GOING, LIKE WHEN MAIA THREW THE ROCKS, BUT NOT BE SURE WHERE HE IS.

SOME OF THESE BOOKS USE PROFESSOR J'S FORMULA, BUT SOME DROP THE 1/2, THE π, OR BOTH, OR OTHERWISE CHANGE THE CONSTANT. HOW DO WE KNOW WHAT TO USE?

MAYBE IT DEPENDS ON HOW UNCERTAINTY IS DEFINED. I BET FOR THESE TASKS WE JUST HAVE TO GET ANSWERS ROUGHLY RIGHT--A FACTOR OF TWO HERE OR THERE DOESN'T CHANGE MUCH.

Terrance is right. Also, some books write $\hbar$ to mean $h/(2\pi)$.

TASK 21.5

Use the fact that a hydrogen atom is approximately 0.1 nm across to estimate the minimum kinetic energy the electron in a hydrogen atom could possibly have. Give your answer in electron volts.

WHY DOES AN ELECTRON HAVE A MINIMUM ENERGY AT ALL?

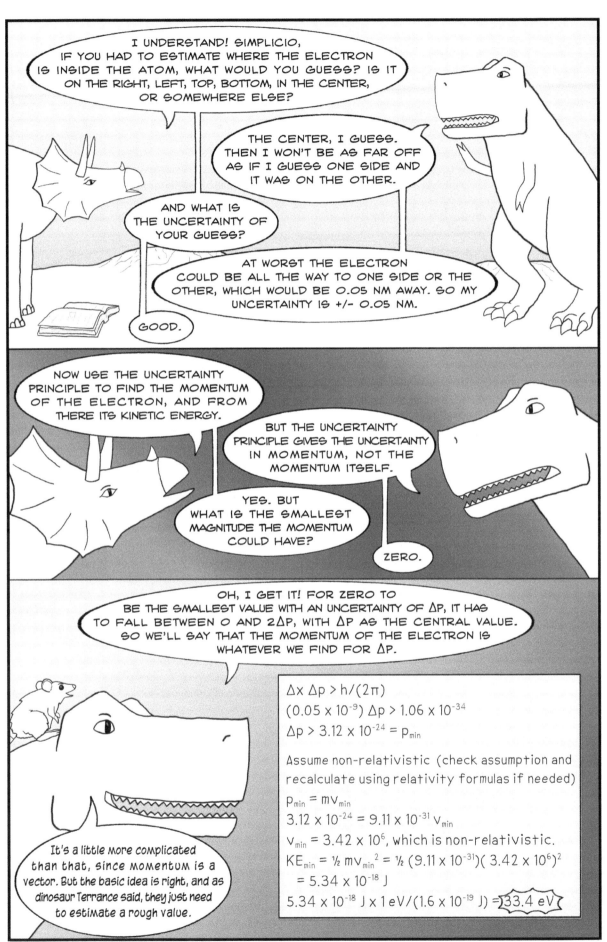

241

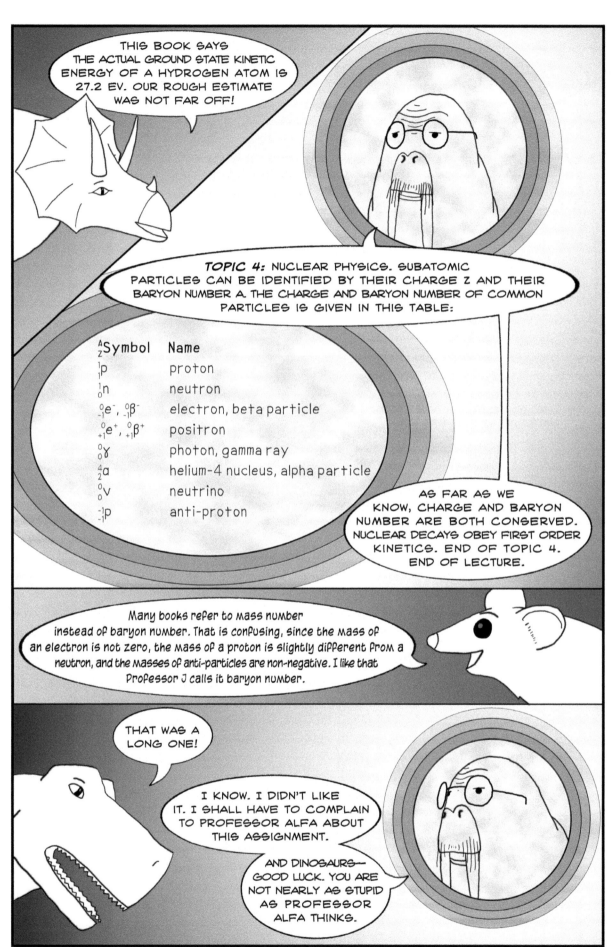

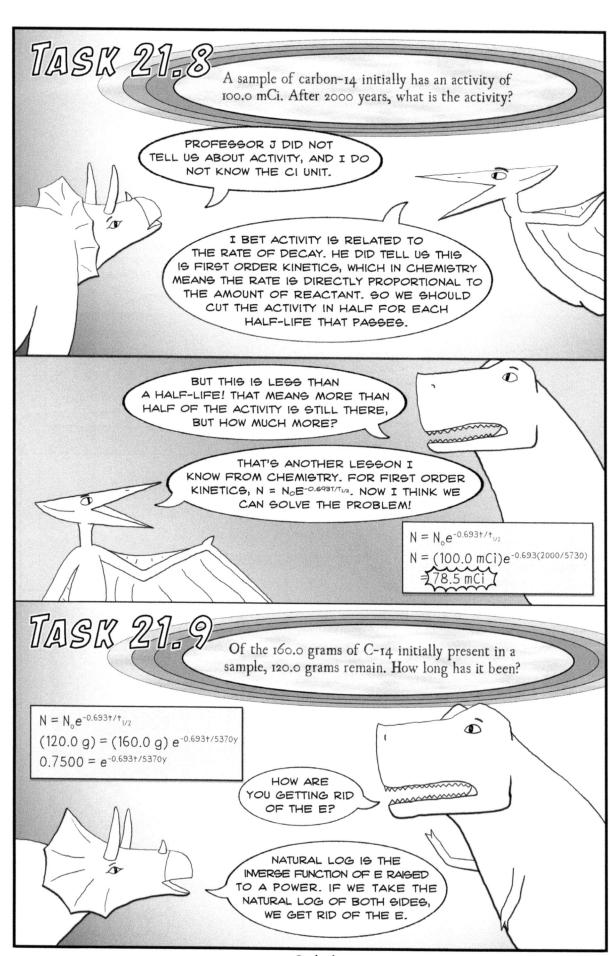

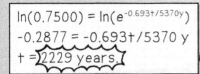

TASK 21.8
A sample of carbon-14 initially has an activity of 100.0 mCi. After 2000 years, what is the activity?
PROFESSOR J DID NOT TELL US ABOUT ACTIVITY, AND I DO NOT KNOW THE CI UNIT.
I BET ACTIVITY IS RELATED TO THE RATE OF DECAY. HE DID TELL US THIS IS FIRST ORDER KINETICS, WHICH IN CHEMISTRY MEANS THE RATE IS DIRECTLY PROPORTIONAL TO THE AMOUNT OF REACTANT. SO WE SHOULD CUT THE ACTIVITY IN HALF FOR EACH HALF-LIFE THAT PASSES.
BUT THIS IS LESS THAN A HALF-LIFE! THAT MEANS MORE THAN HALF OF THE ACTIVITY IS STILL THERE, BUT HOW MUCH MORE?
THAT'S ANOTHER LESSON I KNOW FROM CHEMISTRY. FOR FIRST ORDER KINETICS, $N = N_0E^{-0.693T/T_{1/2}}$. NOW I THINK WE CAN SOLVE THE PROBLEM!
$N = N_0e^{-0.693t/t_{1/2}}$
$N = (100.0\ mCi)e^{-0.693(2000/5730)}$
$= 78.5\ mCi$
TASK 21.9
Of the 160.0 grams of C-14 initially present in a sample, 120.0 grams remain. How long has it been?
$N = N_0e^{-0.693t/t_{1/2}}$
$(120.0\ g) = (160.0\ g)\ e^{-0.693t/5370y}$
$0.7500 = e^{-0.693t/5370y}$
HOW ARE YOU GETTING RID OF THE E?
NATURAL LOG IS THE INVERSE FUNCTION OF E RAISED TO A POWER. IF WE TAKE THE NATURAL LOG OF BOTH SIDES, WE GET RID OF THE E.

$$\ln(0.7500) = \ln(e^{-0.693t/5370y})$$
$$-0.2877 = -0.693t/5370 \text{ y}$$
$$t = 2229 \text{ years.}$$

THAT SEEMS RIGHT. MORE THAN HALF THE SAMPLE IS STILL THERE, SO IT MUST BE LESS THAN A HALF LIFE.

CONGRATULATIONS, DINOSAURS, THAT WAS YOUR FINAL LECTURE.

YOU HAVE DEMONSTRATED THAT, UNDER THE TUTELAGE OF THE WORLD'S GREATEST PHYSICS PROFESSORS, IT IS POSSIBLE TO TEACH DINOSAURS SOME PHYSICS. WE CAN BUILD ON THIS SUCCESS TO SECURE FURTHER LUCRATIVE GRANTS.

THERE'S NO FINAL EXAM?

THERE IS AN EXAM, DINOSAURS, BUT WE WILL NOT BE THE ONES TO ADMINISTER IT. IT WILL INDEED BE FINAL, AND THERE WILL BE NO MAKE-UPS AND NO HELP FROM NAIVE JUNIOR PROFESSORS. REALITY WILL TAKE PRECEDENCE, FOR NATURE CANNOT BE FOOLED.

EQUATIONS IN THIS CHAPTER

$$E_{photon} = hf$$
$$\lambda = h/p$$
$$\Delta x\, \Delta p \geq h/(2\pi$$
$$\Delta E\, \Delta t \geq h/(2\pi)$$
$$N = N_o e^{-0.693t/t_{1/2}}$$

WORDS OF WISDOM

Keep track of which equations apply to photons
and which apply to particles like electrons

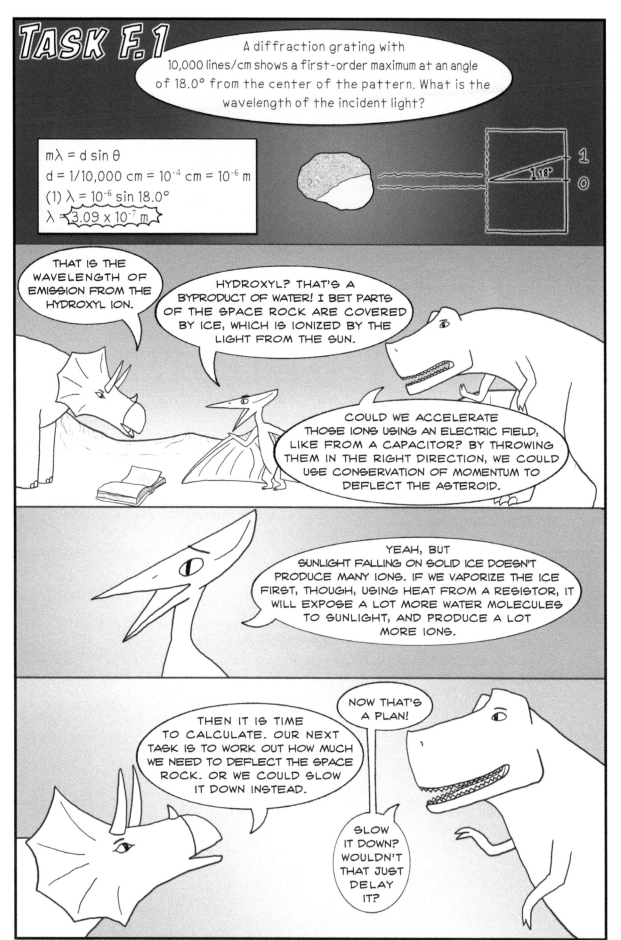

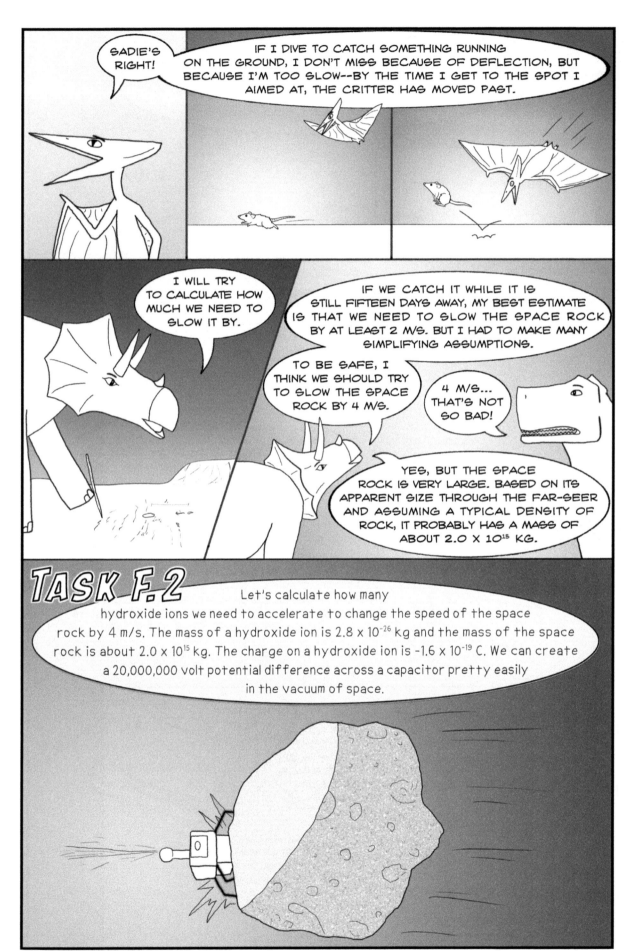

Conservation of energy to find speed of hydroxide ion:

$W_{nc} = E_f - E_i = 0$

Assume the final hydroxide ion is non-relativistic:

$E_f = qV + 1/2\ mv^2$; $E_i = 0$

$0 = qV + 1/2\ mv^2 = (-1.6 \times 10^{-19})(20,000,000) + 1/2\ (2.8 \times 10^{-26})\ v^2$

$v = 1.5 \times 10^7$ m/s, which confirms it is non-relativistic to two significant figures.

Momentum of one ion = $mv = (2.8 \times 10^{-26})(1.5 \times 10^7) = 4.2 \times 10^{-19}$

Momentum of n ions = $n\ (4.2 \times 10^{-19})$

Desired change in momentum of space rock = $(2.0 \times 10^{15})(-4) = -8.0 \times 10^{15}$

Momentum is conserved, so total change in momentum is zero:

$0 = n\ (4.2 \times 10^{-19}) + (-8.0 \times 10^{15})$

$n = 1.9 \times 10^{34}$ ions

TASK F.3

LET'S SEE IF I REMEMBER ENOUGH CHEMISTRY TO CALCULATE HOW MUCH ICE WE NEED TO MAKE THAT MANY HYDROXIDE IONS...GOT IT!

We need at least 5.7×10^8 kg of ice on the space rock. It starts at -150°C, and we need to raise it to 100°C, changing it first to water and then to steam. I can make small electric heaters using a 15,000 V power supply and a 0.40 Ω resistor as a heater. I could fit 50 of those in our rocket. How long would it take for those heaters to do the job? The latent heat of fusion of ice is 3.34×10^5 J/kg, and the latent heat of vaporization of water is 2.26×10^6 J/kg.

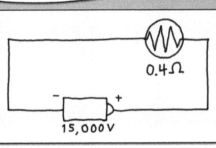

0.4 Ω

15,000 V

Current in circuit:

$V = IR$

$15,000 = I(0.40)$

$I = 3.8 \times 10^4$ A

Power delivered to resistor:

$P = IV = (3.8 \times 10^4)(15,000) = 5.6 \times 10^8$ W

That's for one heater. For 50 heaters,

$P = 50(5.6 \times 10^8) = 2.8 \times 10^{10}$ W

Raise temperature of ice from -150°C to 0°C.

$q = mc\Delta T = (5.7 \times 10^8)(2050)(0 - -150)$

$= 1.8 \times 10^{14}$ J

$t = q/P = 1.8 \times 10^{14}/2.8 \times 10^{10} = 6400$ s.

IT WILL NOT REQUIRE QUITE AS MUCH TIME AS THE DINOSAURS WILL CALCULATE, BECAUSE ICE TURNS TO VAPOR AT A LOWER TEMPERATURE IN THE VACUUM OF SPACE.

BUT IT IS OK IF THEY OVERESTIMATE THE TIME. THEY WILL JUST SLOW THE SPACE ROCK MORE THAN THEY EXPECTED.

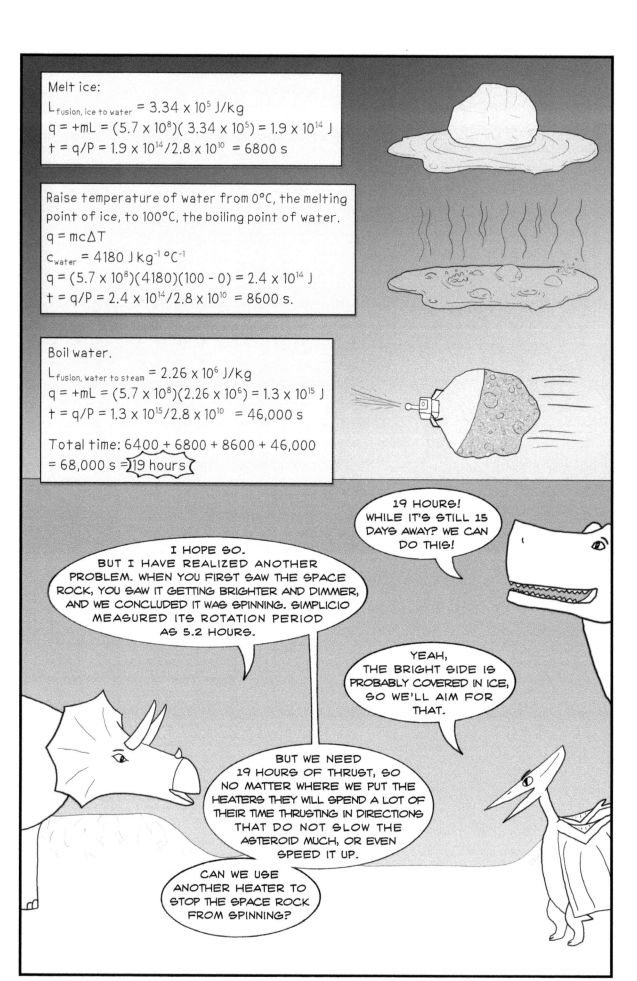

Melt ice:

$L_{fusion, ice to water} = 3.34 \times 10^5$ J/kg

$q = +mL = (5.7 \times 10^8)(3.34 \times 10^5) = 1.9 \times 10^{14}$ J

$t = q/P = 1.9 \times 10^{14}/2.8 \times 10^{10} = 6800$ s

Raise temperature of water from 0°C, the melting point of ice, to 100°C, the boiling point of water.

$q = mc\Delta T$

$c_{water} = 4180$ J kg^{-1} °C^{-1}

$q = (5.7 \times 10^8)(4180)(100 - 0) = 2.4 \times 10^{14}$ J

$t = q/P = 2.4 \times 10^{14}/2.8 \times 10^{10} = 8600$ s.

Boil water.

$L_{fusion, water to steam} = 2.26 \times 10^6$ J/kg

$q = +mL = (5.7 \times 10^8)(2.26 \times 10^6) = 1.3 \times 10^{15}$ J

$t = q/P = 1.3 \times 10^{15}/2.8 \times 10^{10} = 46,000$ s

Total time: 6400 + 6800 + 8600 + 46,000 = 68,000 s = 19 hours

19 HOURS! WHILE IT'S STILL 15 DAYS AWAY? WE CAN DO THIS!

I HOPE SO. BUT I HAVE REALIZED ANOTHER PROBLEM. WHEN YOU FIRST SAW THE SPACE ROCK, YOU SAW IT GETTING BRIGHTER AND DIMMER, AND WE CONCLUDED IT WAS SPINNING. SIMPLICIO MEASURED ITS ROTATION PERIOD AS 5.2 HOURS.

YEAH, THE BRIGHT SIDE IS PROBABLY COVERED IN ICE, SO WE'LL AIM FOR THAT.

BUT WE NEED 19 HOURS OF THRUST, SO NO MATTER WHERE WE PUT THE HEATERS THEY WILL SPEND A LOT OF THEIR TIME THRUSTING IN DIRECTIONS THAT DO NOT SLOW THE ASTEROID MUCH, OR EVEN SPEED IT UP.

CAN WE USE ANOTHER HEATER TO STOP THE SPACE ROCK FROM SPINNING?

Angular momentum of a single ion = $I\omega$

For a particle, $I = mr^2$, where r is the distance of the particle from the center of the circle it makes.

$I_{ion} = (2.8 \times 10^{-26})(5500)^2 = 8.47 \times 10^{-19}$

$v_{ion} = r\omega_{ion}$

$1.5 \times 10^7 = 5500\omega_{ion}$

$\omega_{ion} = 2.7 \times 10^3$ rad/s

$L_{ion} = I_{ion}\omega_{ion} = (8.47 \times 10^{-19})(2.7 \times 10^3) = 2.3 \times 10^{-15}$

For n ions, $L_{all\ ions} = n(2.3 \times 10^{-15})$

We need the angular momentum of the ions to cancel out the angular momentum of the rock, so that it stops spinning.

$8.1 \times 10^{18} - n(2.3 \times 10^{-15}) = 0$

$n = 3.5 \times 10^{33}$ ions

USING CHEMISTRY AGAIN, THAT'S 1.1×10^8 KG OF ICE.

TASK F.5

Calculate the amount of time it would take 4 of the heaters from Task F.3 to melt 1.1×10^8 kg of ice.

WE CAN TAKE A SHORT-CUT. WE KNOW THAT THERE ARE 4 HEATERS INSTEAD OF 50, SO THE POWER IS 4/50 OF TASK F.3. BUT THERE IS ONLY 1.1×10^8 KG OF ICE, INSTEAD OF 5.7×10^8 KG. THE TIME SHOULD BE PROPORTIONAL TO THE MASS OF ICE TO BE MELTED, AND INVERSELY PROPORTIONAL TO THE POWER.

$t = [(1.1 \times 10^8 / 5.7 \times 10^8)/(4/50)]$ 19 hours

= 46 hours.

SO IT WILL TAKE US ABOUT TWO DAYS TO STOP THE SPACE ROCK FROM SPINNING, AND THEN ANOTHER DAY TO SLOW IT ENOUGH TO MISS.

IF WE GET THE ROCKET LAUNCHED SOON ENOUGH, WE'RE OK!

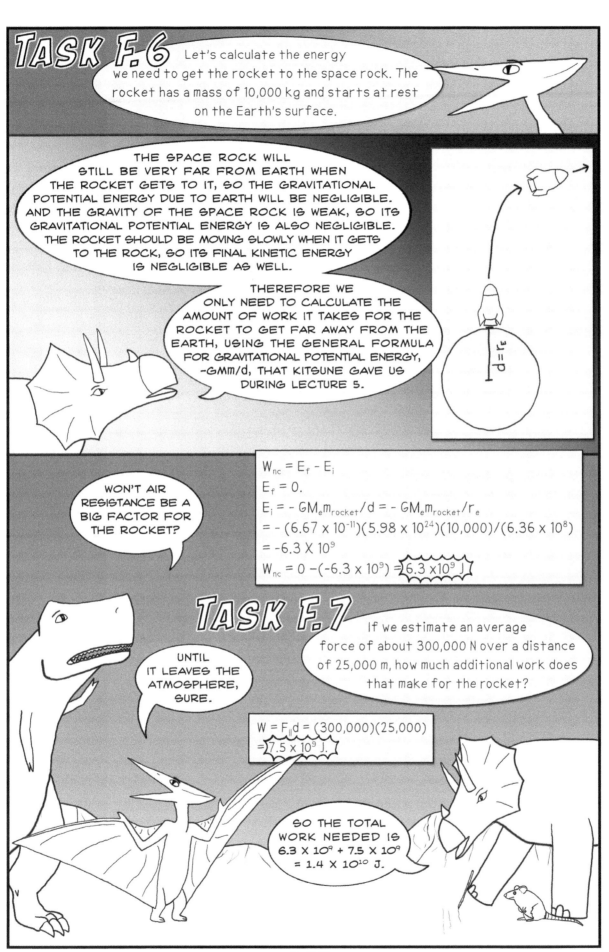

TASK F.6 Let's calculate the energy we need to get the rocket to the space rock. The rocket has a mass of 10,000 kg and starts at rest on the Earth's surface.

THE SPACE ROCK WILL STILL BE VERY FAR FROM EARTH WHEN THE ROCKET GETS TO IT, SO THE GRAVITATIONAL POTENTIAL ENERGY DUE TO EARTH WILL BE NEGLIGIBLE. AND THE GRAVITY OF THE SPACE ROCK IS WEAK, SO ITS GRAVITATIONAL POTENTIAL ENERGY IS ALSO NEGLIGIBLE. THE ROCKET SHOULD BE MOVING SLOWLY WHEN IT GETS TO THE ROCK, SO ITS FINAL KINETIC ENERGY IS NEGLIGIBLE AS WELL.

THEREFORE WE ONLY NEED TO CALCULATE THE AMOUNT OF WORK IT TAKES FOR THE ROCKET TO GET FAR AWAY FROM THE EARTH, USING THE GENERAL FORMULA FOR GRAVITATIONAL POTENTIAL ENERGY, $-GMm/d$, THAT KITSUNE GAVE US DURING LECTURE 5.

$d = r_e$

WON'T AIR RESISTANCE BE A BIG FACTOR FOR THE ROCKET?

$W_{nc} = E_f - E_i$

$E_f = 0.$

$E_i = - GM_e m_{rocket}/d = - GM_e m_{rocket}/r_e$

$= - (6.67 \times 10^{-11})(5.98 \times 10^{24})(10,000)/(6.36 \times 10^8)$

$= -6.3 \times 10^9$

$W_{nc} = 0 - (-6.3 \times 10^9) = 6.3 \times 10^9 \text{ J}$

TASK F.7 If we estimate an average force of about 300,000 N over a distance of 25,000 m, how much additional work does that make for the rocket?

UNTIL IT LEAVES THE ATMOSPHERE, SURE.

$W = F_{\parallel} d = (300,000)(25,000) = 7.5 \times 10^9 \text{ J}.$

SO THE TOTAL WORK NEEDED IS $6.3 \times 10^9 + 7.5 \times 10^9 = 1.4 \times 10^{10}$ J.

253

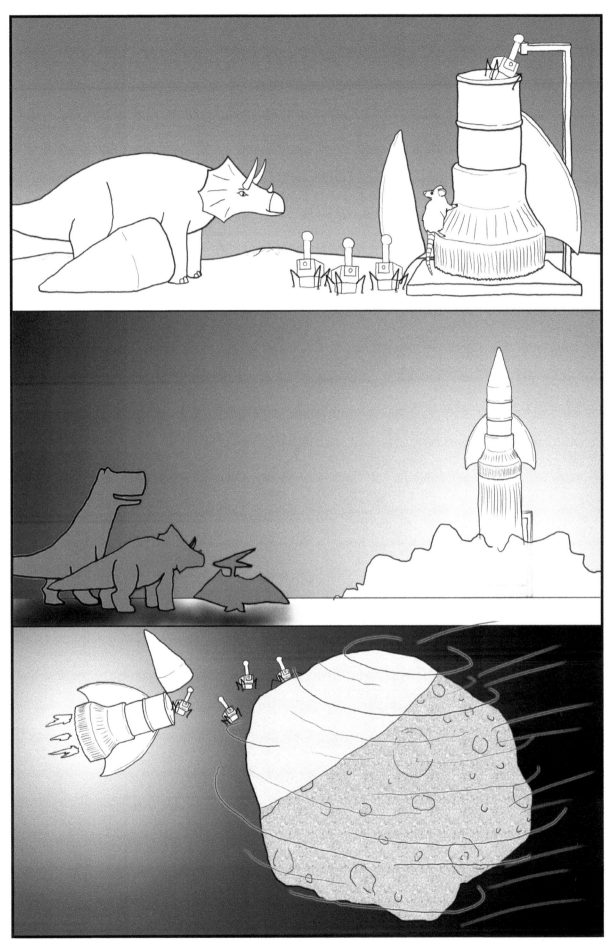

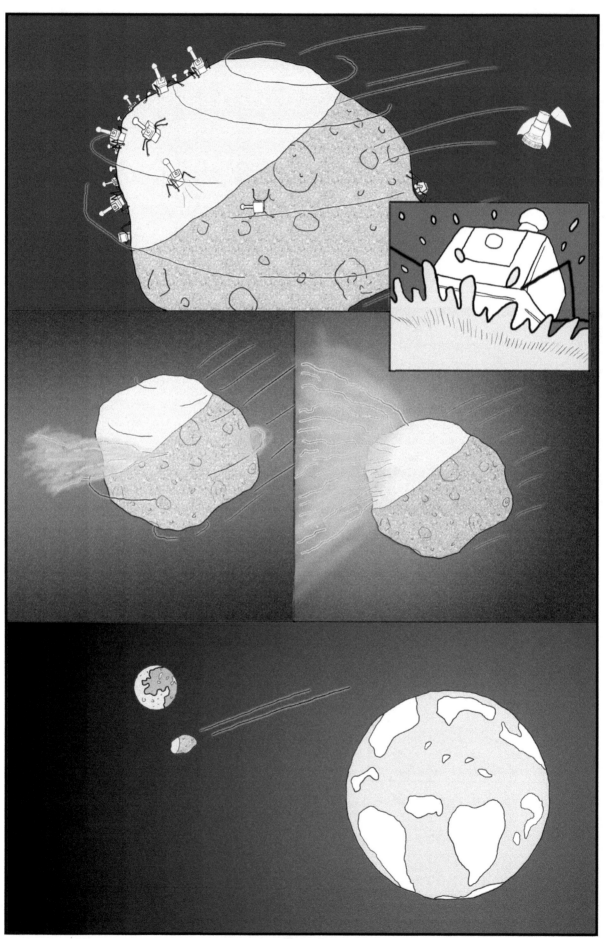

259

FREQUENTLY ASKED QUESTIONS (FAQ)

HOW TO USE THIS BOOK

Do I need to read all these FAQs?
No, although we suggest reading the "Details and Conventions" section.

Do I have to go through this book in order?
No, particularly if you don't care about the plot. You can jump straight to the chapter you need. In some cases, however, you may be encouraged to go back to an earlier chapter to learn a basic technique. Rotational Physics (Chapter 7), for instance, is almost entirely based on material covered earlier. But Energy (Chapter 5) is an example of a chapter that could be done on its own.

Can I use this book to review particular kinds of tasks?
Yes! At the end of the book is a list of "task tags," which are kind of like an index, but with a focus on the skills tested by a particular tag. For example, looking up "energy" will list every task which uses that technique, regardless of the chapter. This makes it easy to review a particular technique.

Why aren't there task tags for objects—like "car problems" or "pulley problems"?
In subjects other than physics, the object mentioned in a question often gives you clues about how you should answer it. In chemistry, for example, discussions of acids will usually mean you should apply the techniques specific to acids. But in physics, the same object can be analyzed many different ways. A problem with a pulley in it could be asking about forces, or energy, or angular momentum. To do well in physics, you must learn to identify the clues that tell you what technique is best to apply, rather than focus on the object being asked about.

How do I learn which technique is best to use with a given task?
That's what Interlude 2 teaches! There's a particularly handy flowchart at the end.

Why does the book have a plot? Why not just show the task solutions?
If physics were just about memorizing a bunch of solutions, that would be fine. But that's not the challenge of physics. Instead, you have to learn how to choose a technique and apply it successfully when faced with a question you have never seen before. On top of that, sometimes physics is not taught well, and, depending on your background and preparation, you might find textbooks confusing at first. So this book aims to show how to go about learning physics when faced with those kinds of challenges. At the start of the plot, the dinosaurs literally can not read the textbooks, and the formal lectures are comically unhelpful. But they demonstrate how to cope when faced with limited knowledge and confusing teaching. Over time, the dinosaurs even learn to appreciate some of the teaching techniques that at first did not work for them. The plot helps us show that in a way that a simple catalogue of solutions would not.

OK, so spill the beans: how do I do well if I'm in a physics class that's taught poorly?

1. According to research, the most important thing is to form a study group, as our dinosaurs (and Maia!) have done here. That's a better predictor of success in physics than anything else, including mathematical preparation, previous physics courses, etc.

2. Find, if you can, a Simplicio, a Sadie, a Terrance, and a Maia. Simplicio asks lots of questions and isn't afraid of trying new things. Sadie has a good background in mathematics and is good at following procedures. Terrance has a good physical sense of the world around him. Maia understands the textbooks better than the others. If you can form a study group with a mix of different skills like that, and value the skills that each of you have, it will help.

3. Find a Kitsune. She might be a TA, or a private tutor, or a family member. But find someone who has been through the course before and is willing and able to help!

4. You don't always have to complete tasks the way the professor does in class. Most professors value effective solutions however they are done, as long as it is possible to follow the logic.

Is there a sample final?

The last chapter serves as a sample final.

DETAILS AND CONVENTIONS

Why don't the step-by-step solutions have units?

Many physics professors encourage students to include units on every step of completing a task. This has the advantage of helping to catch algebraic and careless errors. But it also clutters the page, and can lead to confusion: does m stand for mass, or for meters? Fortunately, physicists have devised a method which allows us to work out physics tasks without including units in every step: the SI system. As long as all quantities are expressed in SI units (see the second page from the front for a list!), then the units will automatically "work out". To use this procedure, make sure to convert all non-SI units to SI units when beginning a task. Then work through the task without units. Finally, add the appropriate SI unit to the final answer. This is the procedure used by many physicists when they are working on their own research, and we use it here. Note: final answers *must* have units, or else they don't mean anything in the real world! You can skip units while completing a task, but not when presenting the final answer.

How do you handle rounding and significant figures?

We try to keep one more significant figure in the details of the solution than was given in the task, and then round back to the number of significant figures which were given at the end. Occasionally, we may keep more digits than that in our actual computation, to avoid rounding errors from building up.

Do you show *every* step of the math?

Not every step, but we try to show enough to let you follow the solutions if you know basic algebra, and we take care to discuss any mathematical concepts, such as trigonometry, that go beyond that.

Why are some expressions in the task solutions grey?
We use grey whenever we substitute an algebraic expression we previously solved for into an equation.

I notice some solutions are labelled with the task number and others aren't. Why is that?
We label solutions with the task number when there is significant intervening discussion between the presentation of the task and its main solution.

Your vector arrows aren't drawn to scale.
That's not a question! No, they're not drawn to scale. One reason is that when you (or our dinosaurs) attempt a task, you don't know the answers in advance, and so don't know how long to draw the vector arrows, and our illustrations reflect that. Another reason is that sometimes drawing them to scale would interfere with clarity.

THE POINT OF THIS BOOK

Who is this book written for?
The primary audience for this book is students in algebra-based college physics classes.

Who else might find this book enjoyable or useful?
Several groups:

1. Those in calculus-based college physics classes. The difficult tasks in calculus-based college physics tend not to be calculus based! Calculus is more a way of using math to do some things more efficiently than it is a way of coming up with harder tasks, at least at the general-physics level. So this book should be useful to them too.
2. Students in challenging high school physics classes. Some high school classes rely mostly on "find the right formula" kinds of questions—this book will not be useful for them. But others, particularly the Advanced Placement sequences, more closely approach this level of task.
3. People who have already taken college physics but are unsatisfied with their skill level or would like to think about the subject in a different way.
4. People who want to self-teach themselves quantitative physics. If you're in this category, note that this book alone is not enough (see the FAQ on "Can I learn physics just from this book?").
5. Fans of unusual graphic novels. It's possible to read this book without understanding all the ins and outs of the individual tasks!

Can I learn physics just from this book?
No. This book teaches how you to solve typical college-level physics tasks effectively. You need to complement it with something that covers the basic concepts more fully. That could, for example, be a traditional textbook or an online course.

Does this book cover everything in algebra-based general physics?
In terms of broad topics, the answer is yes for most courses.

Does a typical algebra-based physics course cover all the topics in this book?
No. Usually some topics get cut out, although what they are vary from course to course.

Does this book cover all the kinds of questions I might see on exams?
No, not directly. This book is focussed on long-form "problem-solving". These kinds of questions form the core of most physics exams, and are what students generally worry most about. But some courses also ask short-answer "conceptual" questions. While some concepts are touched on in this book, it does not focus on those kinds of questions.

I'M CURIOUS...

Why do you call them "tasks" rather than "problems"?
A problem sounds unpleasant, like something you'd rather not have. Task is more neutral. Since some of us enjoy working out these kinds of physics puzzles, we felt completing tasks was a better description than solving problems. We still call the step-by-step details a solution, though, because a completion doesn't sound right.

Have you tried out these tasks with real students?
Yes. Scott Calvin ran a tutoring service for many years. The tasks in this book are a slightly modified version of a set he used in that tutoring service. Over the years, hundreds of students have attempted them.

Are the dinosaurs and professors based on specific people?
No.

Where did you get the names for the dinosaurs and professors?
Simplicio is named after a character Galileo invented for his dialogues on physics. Galileo never said his character wasn't a t-rex, so we went with it.

Terrance is a pteranodon, which sounds like Terrance. We're clever that way.

Sadie was originally going to be named Sarah, both because she is a triceratops and in honor of Sarah Lawrence College, where both authors studied or worked. But we didn't want to confuse her with a certain other triceratops named Cera. So, since Sarah Lawrence's double-secret nickname is Sadie Lou, we went with Sadie.

Maia is an alphadon, which is a cretaceous-era mammal, and thus the ancestor of all of us. (you're a mammal, right?) Maia is Greek for mother.

Kitsune is Japanese for fox. Yes, we know she's a lemur. There are plenty of humans with the last name Fox, so why not a lemur?

Alfa is a variation of alpha, the leader of the pack.

Maxwell is named after James Clerk Maxwell, a famous 19th century physicist who poked fun at bad physics teaching. We like to think, were he still alive today (presumably because he became a vampire or invented a time machine), that he would appreciate our use of satire.

The dinosaurs have technology? They have rulers and rockets and trampolines? They know trigonometry and chemistry? But they don't know what an automobile is?

These dinosaurs do. You're worried about this, but it doesn't bother you we have a talking jaguar in a lab coat?

Are tachyon generators and hyperon signallers real things?

No.

Did you spell gasses wrong?

Some words have multiple acceptable spellings. Many of those involve the question of whether to double a letter: "canceled" or "cancelled"? In this book, we generally opt for the doubled letter when it is allowed.

Who is Scott Calvin?

Professor Scott Calvin (he) is the Pre-Health Program Director at Lehman College of the City University of New York, where he also teaches physics courses. Prior to this, he was a professor of physics at Sarah Lawrence College, where he has developed courses such as Crazy Ideas in Physics, Rocket Science, and Steampunk Physics. He has also taught at the Hayden Planetarium, Examkrackers, the University of San Francisco, and Southern Connecticut State University, as well as operating Mirare Services, a private tutoring agency. He has authored *XAFS for Everyone*, a textbook on x-ray absorption spectroscopy; and *Beyond Curie: Four Women in Physics and Their Remarkable Discoveries, 1903 to 1963*; co-authored *Examkrackers 1001 Questions in MCAT Chemistry*, a best-selling chemistry test-prep book; and co-designed and produced an artisanal pop-up book promoting the National Synchrotron Light Source II facility.

Who is Kirin Furst?

Kirin Emlet Furst (she or they) is an Assistant Professor of Environmental Engineering at George Mason University, where she conducts research on topics in water quality engineering and aquatic chemistry. Professor Furst holds an MS and PhD in Environmental Engineering & Science from Stanford University, and a BA from Sarah Lawrence College where she studied physics under the tutelage of Professor Scott Calvin. Furst is a visual learner and often doodled diagrams and cartoons in her STEM classes to teach herself the material and explain it to others. She is thrilled to share her illustrations and insights with students in college-level physics courses and hopes that this book makes the journey easier and a lot more fun.

Did anyone else work on this book?

Blaine Alleluia provided copy editing, and helped make sure the characters and plot were consistent. Kelsey Monson checked the solutions for accuracy. Kimmie Nguyen provided touch-up art. And Elena Hartley was invaluable in providing additional artwork used throughout the second half of the book.

Have the characters in this book appeared anywhere else?

Simplicio, Kitsune, and the kangaroo Dysnomia appeared in *XAFS for Everyone*, also published by Taylor and Francis. The other characters are original to this book.

Is there anyone else you'd like to acknowledge?

We'd like to thank Lu Han for fruitful discussions during our initial conceptualization of this book, and Dr. Bruce Ravel, a physicist at the National Institute of Standards and Technology for helpful feedback. We'd also like to acknowledge all our supporters on Kickstarter, especially Jamie Duif Calvin, Bruce Ravel, Dorothy and Allen Calvin, and Roberta Carlisle (in memory of Elinor and Harry Emlet).

What fonts did you use in this book?

Webletterer Pro, Blambot Casual, Astounder Squared, NipCen's Print Unicode, UglyQua, VAG-Handwritten, and Chemical Reaction B BRK.

TASK TAGS

BOLD INDICATES THE CONCEPT IS CENTRAL TO SOLVING THE TASK.

Kitsune's Units Crib Sheet

Quantity	Symbol	Unit	SI	Example
position, length, height, etc.	s, x, h, d, r	meter	m	m	tall human: 2 m
mass	m	kilogram	kg	kg	typical human: 80 kg
time, period	t, T	second	s	s	merry-go-round period: 20 s
velocity, speed	v			m/s	highway speed: 25 m/s
acceleration	a			m/s^2	sports car: 5 m/s^2
force, weight	F, W	Newton	N	$kg\,m/s^2$	typical human: 800 N
coefficient of friction	μ			—	running shoes on pavement: 0.9
energy, work, heat	E, W, Q	Joule	J	$kg\,m^2/s^2$	human sprinting: 750 J
power	P	Watt	W	$kg\,m^2/s^3$	human climbing stairs: 1200 W
spring constant	k			kg/s^2	bedspring: 10,000 N/m
momentum, impulse	p, J			$kg\,m/s$	human sprinting: 300 kg m/s
pressure	P	Pascal	Pa	$kg/(m\,s^2)$	high heel on toe: 6,000,000 Pa
frequency	f	Hertz	Hz	1/s	merry-go-round: 0.05 Hz
angular displacement	θ	radian	rad	rad	right angle: 1.7 rads
angular speed or frequency	ω			rad/s	merry-go-round: 0.3 rad/s
angular acceleration	α			rad/s^2	baseball bat: 50 rad/s^2
torque	τ		N m	$kg\,m^2/s^2$	opening a door: 200 N m
moment of inertia	I			$kg\,m^2$	frisbee: 0.001 $kg\,m^2$
angular momentum	L			$kg\,m^2/s$	frisbee: 0.05 $kg\,m^2/s$
density	ρ			kg/m^3	water: 1000 kg/m^3
area	A			m^2	face of a quarter: 0.0005 m^2
volume	V			m^3	typical human: 0.08 m^3
temperature	T	Kelvin, Celsius	K, °C	K	warm day: 300 K, 27 °C
specific heat	c		J/(kg °C)	$m^2/(s^2\,K)$	water: 4180 J/(kg °C)
latent heat	L		J/kg	m^2/s^2	water vaporization: 2,260,000 J/kg
entropy	S		J/K	$kg\,m^2/(s^2\,K)$	pot of boiling water to steam: 6060 J/kg
charge	q, Q	Coulomb	C	A s	typical static: 1 x 10^{-9} C
electric field	E		V/m or N/C	$kg\,m/(s^3\,A)$	spark in dry air: 3 x 10^6 V/m
electric potential, EMF	V, $\mathcal{E}$	Volt	V	$kg\,m^2/(s^3\,A)$	9-V battery
magnetic field	B	Tesla	T	$kg/(s^2\,A)$	Earth's field: 0.00005 T
current	I	Ampere	A	A	light bulb: 0.5 A
resistance	R	Ohm	Ω	$kg\,m^2/(s^3\,A^2)$	light bulb: 240 Ω
resistivity	ρ		Ω m	$kg\,m^3/(s^3\,A^2)$	copper: 1.7 x 10^{-8} Ω m
capacitance	C	Farad	F	$s^4\,A^2/(kg\,m^2)$	typical: 1 x 10^{-9} F
dielectric constant	κ			—	glass: 6
magnetic flux	Φ	Weber	Wb	$kg\,m^2/(s^2\,A)$	Earth's field through human: 5 x 10^{-5} Wb
inductance	L, M	Henry	H	$kg\,m^2/(s^2\,A^2)$	bedspring: 0.001 H
optical power	P	diopter	D	1/m	reading glasses: 3 D
intensity	I		W/m^2	kg/s^3	sunlight: 1000 W/m^2

Metric Prefixes

c	centi	10^{-2}		k	kilo	10^3
m	milli	10^{-3}		M	mega	10^6
μ	micro	10^{-6}		G	giga	10^9
n	nano	10^{-9}				
p	pico	10^{-12}				

Simplicio's Formula Sheet

Kinematics
(Constant acceleration)
$$v = v_0 + at$$
$$\Delta x = v_0 t + \tfrac{1}{2} at^2$$

Dynamics
$$\sum F = ma$$
$$F_{friction}/F_{normal} \leq \mu$$
$$F_B = \text{weight of fluid displaced}$$
$$F_{spring} = kx$$

Work, Energy, and Heat
$$W_{nc} = E_f - E_i$$
$$W = Fd\cos\theta$$
$$KE = 1/2\, mv^2$$
$$PE_{grav} = mgh$$
$$PE_{elastic} = 1/2\, kx^2$$
$$P = W/t$$
$$P + 1/2\, \rho v^2 + \rho gh = \text{constant}$$
$$Q_{in,\,net} + W_{on,\,net} = \Delta E$$
$$q = mc\Delta T$$
$$q_{phase\ change} = \pm mL$$
$$W_{by\ a\ gas} = P\Delta V$$

Circular Motion
(at constant speed)
$$a_{centripetal} = v^2/r$$
$$f = v/(2\pi r)$$
$$f = 1/\text{Period}$$

Odds and Ends
$$\rho = m/V$$
$$P = F/A$$
$$\Delta S \geq Q/T$$
Incompressible fluid:
$$vA = \text{constant}$$

Quantum Physics
$$E_{photon} = hf$$
$$\lambda = h/p$$
$$\Delta x\, \Delta p \geq h/(2\pi)$$
$$\Delta E\, \Delta t \geq h/(2\pi)$$
$$N = N_0 e^{-0.693t/t_{1/2}}$$

Linear		Angular	Bridge
s	θ	angular displacement	$s = r\theta$
v	ω	angular speed	$v_t = r\omega$
a_t	α	angular acceleration	$a_t = r\alpha$
t	t	time	t
F	τ	torque	$\tau = rF\sin\theta$
m	I	moment of inertia	
p	L	angular momentum	L
KE	KE	rotational kinetic energy	KE

Impulse-Momentum
$$J_{external} = \Delta p$$
$$J = Ft$$
$$p = mv$$

Waves
$$v = \lambda f$$
Standing, ends same: $f = nv/(2L)$ $n = 1, 2, 3, \ldots$
Standing, ends different: $f = nv/(4L)$ $n = 1, 3, 5, \ldots$
Peak $= \sqrt{2}$ (rms)
$$I = S = P/A$$
Sounds intensity level: $\beta_2 - \beta_1 = 10\log(I_2/I_1)$
EM Waves: $E = cB$, $u = \epsilon_0 E^2$, $S = uc$
Interference pattern: $d\sin(\theta) = m\lambda$; $x = L\tan(\theta)$
Double slit/diffraction grating:
$m = 0, 1, 2, \ldots$ gives maxima; d is spacing between slits
Single slit:
$m = 0, 1/2, 3/2, \ldots$ gives maxima; d is width of slit
Unpolarized incident light: $S_{out} = \tfrac{1}{2} S_{in}$
Malus's Law: $S_{out} = S_{in}\cos^2(\theta)$

Simple Harmonic Motion
$$\Delta x = A\cos(\omega t)$$
$$\omega_{mass\ on\ spring} = \sqrt{(k/m)}$$
$$\omega_{pendulum} = \sqrt{(g/L)}$$

Geometrical Optics
Reflection: $\theta_r = \theta_i$
Refraction: $n_1\sin\theta_1 = n_2\sin\theta_2$; $n = c/v$
$$1/f = 1/d_i + 1/d_o; \quad m = h_i/h_o = -d_i/d_o$$
f + for converging lenses and concave mirrors;
f is - for diverging lenses and convex mirrors
d_i is + for a real image; -for a virtual image
d_o and h_o are always + for a single lens or mirror
m is > 1 for an enlarged image; < 1 for a reduced image
h_i is + for an upright image; - for an inverted image

	Resistors	Capacitors	Inductors
In series	Add	$1/C = 1/C_1 + 1/C_2$	Add
In parallel	$1/R = 1/R_1 + 1/R_2$	Add	$1/L = 1/L_1 + 1/L_2$
Voltage	$\Delta V = IR$	$Q = C\Delta V$	$\Delta V = -L\Delta I/\Delta t$
Geometry	$R = \rho L/A$	$Q = \kappa\epsilon_0 A/d$ (parallel-plate)	$L = \mu AN^2/\text{length}$ (solenoid)
Energy	$P = I\Delta V$	Energy $= 1/2\,C(\Delta V)^2$	Energy $= 1/2\,LI^2$

Induction
$$\Phi = B_\perp A = BA\cos\theta$$
$$|EMF| = |N\Delta\Phi/\Delta t|$$
Induced current opposes
the change in flux

Time-Dependent Circuits
RC Circuit: $\tau = RC$
LR Circuit: $\tau = L/R$
Start at max, drop to zero: multiply by $e^{-t/\tau}$
Start at zero, rise to max: multiply by $(1 - e^{-t/\tau})$

Special Relativity
$$v_{AC} = (v_{AB} + v_{BC})/(1 + v_{AB}v_{BC}/c^2)$$
$$v_{BA} = -v_{AB}$$
$$\gamma = 1/\sqrt{1 - v^2/c^2}$$
$$t = \gamma t_0; \quad L = L_0/\gamma$$
$$F = \gamma ma; \quad E = \gamma mc^2$$